Mathematical Neuroscience

Mathematical Neuroscience

Stanisław Brzychczy
Faculty of Applied Mathematics
AGH University of Science and Technology
Kraków, Poland

Roman R. Poznanski
Computation and Informatics Research Cluster
Institute of Research Management and Monitoring
University of Malaya
Kuala Lumpur, Malaysia

AMSTERDAM • BOSTON • HEIDELBERG • LONDON • NEW YORK
OXFORD • PARIS • SAN DIEGO • SAN FRANCISCO • SINGAPORE
SYDNEY • TOKYO
Academic Press is an imprint of Elsevier

Academic Press is an imprint of Elsevier
32 Jamestown Road, London NW1 7BY, UK
225 Wyman Street, Waltham, MA 02451, USA
525 B Street, Suite 1800, San Diego, CA 92101-4495, USA

Notice
No responsibility is assumed by the publisher for any injury and/or damage to persons or property as a matter of products liability, negligence or otherwise, or from any use or operation of any methods, products, instructions or ideas contained in the material herein. Because of rapid advances in the medical sciences, in particular, independent verification of diagnoses and drug dosages should be made.

Library of Congress Cataloging-in-Publication Data
A catalog record for this book is available from the Library of Congress

British Library Cataloguing-in-Publication Data
A catalogue record for this book is available from the British Library

ISBN: 978-0-12-411468-5

For information on all Academic Press publications
visit our website at elsevierdirect.com

Typeset by Scientific Publishing Services (P) Ltd., Chennai
www.sps.co.in

Contents

About the Authors

Stanisław Brzychczy is a Professor of Mathematics at the Faculty of Applied Mathematics at the AGH University of Science and Technology, Kraków. His scholarly interests include partial differential equations and their application to real-world problems. He is a graduate of the Faculty of Mathematics, Physics and Chemistry at the Jagiellonian University, Kraków. He received his Ph.D. (1964) and D.Sc. (1995) degrees in mathematics from the Jagiellonian University. He is a member of the Polish Mathematical Society, the American Mathematical Society, and the Society for Industrial and Applied Mathematics.

Roman R. Poznanski is Chief Editor of the *Journal of Integrative Neuroscience*. He is a graduate of the School of Biological Sciences at the Australian National University (ANU), Canberra. He received his Ph.D. (1992) degree in neuroscience from the ANU. He has edited several contemporary books including *Biophysical Neural Networks* (2001) and *Modeling in the Neurosciences* (2005).

Foreword

In the second decade of the twenty-first century, brain researchers (neuroscientists) have begun to decipher the dynamics of large-scale neural networks and to gauge how the functioning of the brain is dependent on the spatiotemporal integration of the resultant dynamics. Knowing more detailed facts about brain connectivity, we are faced with the problem of how such information can be put together in one system to understand the whole brain and the effective tracing of cause-and-effect relationships determining its actions. Data gathering by neuroinformaticians steadfastly produces a 'brain in a supercomputer' virtual model that can be regularly updated to fill in the missing pieces of the puzzle. Unfortunately, such data gathering does not imply that there is a "glue" that allows us to combine multiple empirical observations into a single theory of the brain.

Experience with the development of other empirical sciences, especially astronomy, physics, and more recently chemistry, shows that the most effective and most efficient "glue" that allows us to combine multiple empirical observations into a single theory is mathematics. Therefore, it is now a very important and urgent task to create the foundations of a new science that has not yet been formulated, but already has a name: Mathematical Neuroscience. It is worth mentioning that when this area is finally formulated, it will be supplied with empirical facts collected from many other fields of study related to the brain, but also the behavior of humans and animals, so its development on such a rich database of empirical facts should be very fruitful.

In my opinion, *Mathematical Neuroscience,* for which I have the honor and pleasure of writing this foreword, is a step in the right direction. Creating a mathematical description of neural systems is very important and necessary. Of course, there are a lot of such descriptions and they are well described in the literature. However, what stands out when reviewing all the studies included in mathematical or computational neuroscience is that there is relatively little work of professional mathematicians. In this context, the book presented here is a glorious exception. It presents mathematical descriptions of neurodynamic structures, starting not from the empirical description of the neurobiological processes, but from the mathematical description of a more general class of dynamical systems for which models of neural structures are a special case.

The book begins with a description of methods of functional analysis for infinite systems of equations of the reaction-diffusion type with Volterra functionals. Models such as the monotone iterative technique are defined and tested in order to solve them, and then a discussion is conducted on lower and upper solutions and the truncation method and fixed point method. As well as discussing the methods of solving the considered class of problems, an analysis of stability of solutions is presented. Part II starts with a consideration of the new directions of research that will have the most impact on neuroscience. This includes advances in the analysis of infinite systems of continuous and discrete models of neural systems in infinite-dimensional abstract spaces and neural models in infinite-dimensional spaces and their finite-dimensional projections. There are no other studies in which the problems arising from neuroscience are analyzed with such an advanced level of mathematics and with the use of such a wide variety of tools and mathematical methods.

Professor Ryszard Tadeusiewicz
Head of Biocybernetic Laboratory
AGH University of Science and Technology
Kraków, Poland

Preface

Mathematical Neuroscience presents methods of nonlinear functional analysis and its application to neuroscience. It is the first book to present a compendium of methods of nonlinear analysis to better understand the dynamics associated with solutions of infinite systems of equations.

Application of a range of methods for nonlinear analysis broadly construed in neuroscience is needed because there is no attempt yet for an expansive nonlinear "tool kit" that can be efficiently applied to solve nonlinear problems arising in neuroscience. Applying methods of functional analysis to describe the solutions of nonlinear problems reflects a paradigm change in how to analyze nonlinear problems.

Nonlinear analysis facilitates qualitative solutions of infinite systems of equations in infinite-dimensional space that are considered to be representations of models in neuroscience together with their finite-dimensional projections. This book investigates the solvability of infinite systems of equations with the use of methods on differential inequalities (e.g., comparison theorems, aximum principles), different monotone iterative techniques (e.g., methods of lower and upper solutions), a truncation method, and a topological fixed point method. It also presents a fundamental theory of finite systems of equations with extremely complex forms of nonlinearity.

This book would be suitable as a text material for a one semester course in mathematical neuroscience. It is a must read for graduate students in mathematics who have no prior knowledge of neuroscience since the methods presented provide tools for better understanding. The book might also be of interest for non-mathematical specialists looking for novel mathematical tools not commonly extrapolated in problems arising in neuroscience or as a methodological handbook for those who wish to analyze infinite systems of partial differential equations using methods of nonlinear functional analysis. Intrinsically, this book is a unique and innovative advanced reference work on methods of nonlinear analysis of infinite systems and their application in neuroscience. It can be useful for readers who want to be come acquainted with methods of nonlinear analysis. It is the first reference work on mathematical neuroscience distinguishing itself from computational neuroscience by presenting extremely cogent theory and methods of nonlinear analysis in a single volume.

Preface

This book would never have been written without the professional support of many colleagues who reviewed the manuscript and whose comments were instrumental in making it a success. Our special gratitude goes to Lech Górniewicz, Zdzislaw Jackiewicz, the late Zdzisław Kamont, Henryk Leszczyński, Bolesław Szafirski, and Ryszard Tadeusiwiscz. Each added much to this work, generously providing ideas and remarkable suggestions for improvement. We also appreciate the assistance provided by Daniel Delimita for preparing the LATEX2e version of the handwritten manuscript and EasyEnglishPublication.com for help with producing the final LATEX2e version. Finally, we want to thank Ghuath Jasmon, Vice-Chancellor of the University of Malaya for his generous support.

Stanisław Brzychczy and Roman R. Poznanski

Part I

Methods of Nonlinear Analysis

CHAPTER 1

Introduction to Part I

Part I is devoted exclusively to the theory and methods of nonlinear analysis of infinite systems. These infinite systems play an important role in the integrative aspects of neuroscience modelling with functional analytic methods of describing nonlinear phenomena through set theory and vector spaces. In modern mathematics vector spaces usually are carried over from Cartesian spaces, and knowledge of these spaces is of help in analyzing more general vector spaces (e.g., Banach spaces, Hilbert spaces). However, when it comes to the brain and uncovering its dynamical nature, a vector or point in ordinary three-dimensional space no longer makes any sense. Visualizing neuronal dynamics based on ordinary three-dimensional Cartesian space becomes counterproductive, especially in the process of integrating across scale. The dynamics of the brain with its enormous complexity cannot be understood through the use of elementary functions of classical mathematics. Firstly, we need a broader definition of function so that a noncommittal notation and the nature of the rule or mapping become irrelevant. Secondly, we need to consider spaces of continuous functions such as infinite-dimensional Euclidean vector spaces that carry nonlinear phenomena from which new ideas can be harnessed from dynamical continuity.

The noncommittal approach underlying the foundational theory of infinite systems enables new constructs of more powerful models, especially in the process of integrating across scale. For instance, the set of states of biological neural networks is infinite, topologically continuous. This is because the state of the network is specified not by the firing of digital spikes in neurons, but in the electrical patterns brought about by the electrical field. However, the set of states of artificial neural networks is finite, topologically discrete due to spiking being the predominant paradigm where refractory properties of neurons are included after they have fired.

The methods of nonlinear analysis presented in the book apply only to second-order equations arising in neuroscience. Given the importance of more realistic neuronal models at the cellular level, it can be said that neuronal models are foundational units of the nervous system that can be interconnected to form dendritic trees, networks of dendritic trees, systems of networks, and so on. Consider a tapering cable of a dendritic segment with a continuously tapering cable as shown in Figure 1.1.

Mathematical Neuroscience. http://dx.doi.org/10.1016/B978-0-12-411468-5.00001-6

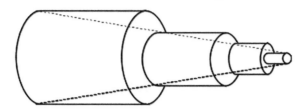

Figure 1.1 A tapering cable representing a single branch of a dendritic tree.

The equation describing the depolarization $V = V(x, t)$ (in millivolts) at the space point x and time t for a continuous radius of the cylinder $r(x)$ denoted by the dashed lines is given by

$$C_m \frac{\partial V}{\partial t} = \frac{1}{2\rho r(x)\sqrt{1 + (dr(x)/dx)^2}} \frac{\partial V}{\partial x}\left(r^2(x)\frac{\partial V}{\partial x}\right) - \sum g_i(x, t; V)(V - V_i). \qquad (1.1)$$

In essence, by assuming $C_m = 1 \ \mu\text{F/cm}^2$, $z(t, x) = V(x, t)$, and the operators $D_t = \frac{\partial V}{\partial t}$, $D_x = \frac{\partial V}{\partial x}$, $D_{xx}^2 = \frac{\partial^2 V}{\partial x^2}$, etc., the above equations turn out to be

$$D_t z(t, x) - \frac{r(x)}{2\rho\sqrt{1 + (dr(x)/dx)^2}} D_{xx}^2 z - \frac{(dr(x)/dx)}{\rho\sqrt{1 + (dr(x)/dx)^2}}$$

$$D_x z - \sum g_i(x, t; z)(z(t, x) - z_i).$$

Define $a(x) = \frac{r(x)}{2\rho\sqrt{1+(dr(x)/dx)^2}}$, $b(x) = \frac{-(dr(x)/dx)}{\rho\sqrt{1+(dr(x)/dx)^2}}$, and $f(t, x, z(t, x), z) = \sum g_i(t, x, z)(z(t, x) - z_i)$ so that the above equation becomes

$$D_t z(t, x) - a(x)D_{xx}^2 z(t, x) + b(x)D_x z(x, t) = f(t, x, z(t, x), z). \qquad (1.2)$$

We consider coupled infinite systems of equations where the coupling arises from both the connectivity between branches of a dendritic tree of a single neuron and the synaptic coupling between individual neurons in networks. We therefore introduce a superscript "i" to denote, for example, the membrane depolarization of the ith neuron where "i" ranges from 1 to ∞. The coefficients $a(x)$ and $b(x)$ have introduced subscripts "j" and "k" to denote the electrogeometrical inhomogeneity of the branching processes and we assume there are "m" branches in a particular dendritic tree, so that (1.2) becomes

$$D_t z^i(t, x) - \sum_{j,k=1}^m a_{jk}(x)D_{x_j x_k}^2 z^i(t, x) + \sum_{j=1}^m b_j(x)D_{x_j} z^i(x, t) = f^i(t, x, z(t, x), z). \qquad (1.3)$$

In set theory notation (1.3) is considered a weakly coupled infinite system of equations of the form

$$\mathcal{F}^i[z^i](t, x) := \mathcal{D}_t z^i(t, x) - \sum_{j,k=1}^m a^i_{jk}(t, x)\mathcal{D}^2_{x_j x_k} z^i(t, x)$$

$$+ \sum_{j=1}^m b^i_j(t, x)\mathcal{D}_{x_j} z^i(t, x) = f^i(t, x, z(t, x), z) \quad \text{for } i \in S \quad (1.4)$$

for $(t, x) = (t, x_1, \ldots, x_m) \in D$, where D is a bounded cylindrical domain, $D := (0, T] \times G$, $0 < T < \infty$, $G \subset \mathbb{R}^m$ is an open and bounded domain whose boundary ∂G is an $(m-1)$-dimensional sufficiently smooth surface $m = 1, 2,$ or 3, S is an arbitrary set of indices, $\mathcal{F}^i$, $i \in S$, are diagonal operators[1] which are uniformly parabolic in $\overline{D}$, and f^i, $i \in S$, are given functions

$$f^i : \overline{D} \times \mathcal{B}(S) \times C_S(\overline{D}) \to \mathbb{R}, \quad (t, x, y, s) \mapsto f^i(t, x, y, s), \quad i \in S,$$

where $S_0 := \{(t, x) : t = 0, x \in \overline{G}\}$, $\sigma := (0, T] \times \partial G$ is the lateral surface and $\Gamma := S_0 \cup \sigma$ is the parabolic boundary of the domain D and $\overline{D} := D \cup \Gamma = [0, T] \times \overline{G}$. The right-hand sides f^i of the system (1.4) are functionals with respect to the last variable. They describe the local nonlinearity due to the membrane electrical properties, and the diffusive interaction due to electrical current flow in neurons.

The following examples of Volterra functionals are considered:

$$f_1(t, x, z) = \int_0^t m(t - \tau)K(z(\tau, x))d\tau, \quad (1.5)$$

$$f_2(t, x, z) = \int_0^t K(t, \tau, x, z(\tau, x))d\tau, \quad (1.6)$$

$$f_3(t, x, z) = \int_0^t \int_G K(t, \tau, x, \xi, z(\tau, \xi))d\tau d\xi. \quad (1.7)$$

Equations including such functionals need to be considered in appropriately chosen domains. Other examples include:

$$f_4^1(t, x, z) = -z^1(t, x)\sum_{k=1}^\infty a_k^1 z^k(t, x) + \sum_{k=1}^\infty b_k^1 z^{1+k}(t, x), \quad (1.8)$$

---

[1] Operators $\mathcal{F}^i$, $i \in S$, are diagonal if $\mathcal{F}^i$ depends on z^i only for all $i \in S$.

$$f_5^j(t, x, z) = \frac{1}{2} \sum_{k=1}^{j-1} a_k^{j-k} z^{j-k}(t, x) z^k(t, x) - z^j(t, x) \sum_{k=1}^{\infty} a_k^j z^k(t, x)$$

$$+ \sum_{k=1}^{\infty} b_k^j z^{j+k}(t, x) - \frac{1}{2} z^j(t, x) \sum_{k=1}^{j-1} b_k^{j-k} \quad \text{for } j = 2, 3 \dots, \tag{1.9}$$

and

$$f_6^1(t, x, z) = -z^1(t, x) \sum_{k=1}^{\infty} \int_G a_k^1(x, \xi) z^k(t, \xi) d\xi$$

$$+ \sum_{k=1}^{\infty} \int_G B_k^1(x, \xi) z^{1+k}(t, \xi) d\xi, \tag{1.10}$$

$$f_7^j(t, x, z) = \frac{1}{2} \sum_{k=1}^{j-1} \int_{G \times G} A_k^{j-k}(x, \xi, \eta) z^{j-k}(t, \xi) z^k(t, \eta) d\xi d\eta$$

$$- z^j(t, x) \sum_{k=1}^{\infty} \int_G a_k^j(x, \xi) z^k(t, \xi) d\xi + \sum_{k=1}^{\infty} \int_G B_k^j(x, \xi) z^{j+k}(t, \xi) d\xi$$

$$- \frac{1}{2} z^j(t, x) \sum_{k=1}^{j-1} b_k^{j-k}(x) \quad \text{for } j = 2, 3, \dots, \tag{1.11}$$

where $\int_G A_k^j(x, \xi, \eta) dx = a_k^j(\xi, \eta)$ and $\int_G B_k^j(x, \xi) dx = b_k^j(\xi)$ are the nonnegative coefficients of rates a_k^j and b_k^j, while f_4^j, f_5^j, f_6^j, and f_7^j are Volterra functionals.

Systems with right-hand sides of this type are considered as the discrete coagulation-fragmentation models with diffusion. There are two different (discrete and continuous) models which describe changes in the concentration of clusters (proteins) undergoing coagulation (composition) or fragmentation (decomposition). If a variable $z^j = z^j(t, x)$ describes the concentration of j clusters comprising j identical basic elements known as monoclusters, then the variable is a discrete one, taking nonnegative integer values. The model thus obtained is a discrete coagulation-fragmentation model built of infinite countable systems of equations.

These models are expressed in terms of the infinite countable systems of equations of the form

$$\frac{\partial z^j(t, x)}{\partial t} - d_j \sum_{l=1}^{m} \frac{\partial^2 z^j(t, x)}{\partial x_l^2} = f_{5,6}^j(t, x, z), \quad j \in \mathbb{N}, \tag{1.12}$$

for $(t, x) \in (0, T] \times G := D$, where $G \subset \mathbb{R}^m$ is a bounded domain with sufficiently smooth boundary ∂G, with the initial condition

$$z(0, x) = U_0(x) \geq 0 \quad \text{for } (t, x) \in G, \tag{1.13}$$

and the sealed-end boundary condition

$$\frac{dz^j(t, x)}{dv} = 0 \quad \text{for } (t, x) \in \sigma, \quad j \in \mathbb{N}, \tag{1.14}$$

where functions f_4^j, f_5^j are given by (1.8), (1.9), respectively, and diffusion coefficients d_j are positive.

This model is expressed in terms of the infinite countable systems of equations of the form

$$\frac{\partial z^j(t, x)}{\partial t} - \mathcal{A}^j[z^j](t, x) = f_{6,7}^j(t, x, z) \quad \text{for } (t, x) \in D \quad j \in \mathbb{N} \tag{1.15}$$

with initial and boundary conditions (1.13) and (1.14), where the diffusion operators, having the divergence form

$$\mathcal{A}^j[z^j](t, x) := \sum_{l,k=1}^{m} \frac{\partial}{\partial x_j} \left(d_{lk}^j(t, x) \frac{\partial z^j(t, x)}{\partial x_k} \right), \quad j \in \mathbb{N}$$

are uniformly elliptic in $\overline{D}$, diffusion coefficients d_{jk}^i are positive and functions f_6^i, f_7^i are given by (1.10), (1.11), respectively, integral operators.

Unfortunately, the monotone iterative method does not cover the systems (1.12) and (1.15). Therefore, to solve problems (1.12) and (1.15), the truncation method is applied.

Infinite countable systems of ordinary differential equations can be used to solve some problems for nonlinear equations with the use of the finite-difference method. In this very important method which is called the numerical method of lines, spatial derivatives only are discretized by difference expressions, leading to an infinite countable system of equations.

On the other hand, when a continuous variable (e.g., concentration per unit volume) taking nonnegative real values is used in the description, we arrive at a continuous model of coagulation-fragmentation, a model built of infinite uncountable systems of equations. Adopting an important assumption that the number of created clusters is not bounded, we conclude that the numbers of both equations and variables may be infinite.

Continuous models may be expressed in terms of infinite uncountable systems of equations of the reaction-diffusion type. Studying infinite uncountable systems of equations is far from simple. In particular, if we have an infinite system of equations of the form (1.4) (nonstationary reaction-diffusion equations), then the study may be brought down to an evolution equation of the form

$$\frac{\partial u}{\partial t} - A(t)u = F(t, u) \quad \text{for } t > 0, \quad x \in \mathbb{R}^m$$

with the initial condition

$$u(0, x) = f(0, x) \quad \text{for } x \in G,$$

where $-A$ is a linear operator and F is a nonlinear reaction term, considered in an appropriately defined vector space.

For system (1.4), we will consider the so-called Fourier first initial-boundary value problem: find the regular (classical) solution z of system (1.4) in $\overline{D}$ fulfilling the initial-boundary condition

$$z(t, x) = \phi(t, x) \quad \text{for } (t, x) \in \Gamma. \tag{1.16}$$

We remark that it is not possible to solve infinite systems of equations. Therefore, a natural approach to the study of the existence and uniqueness of solutions of initial and boundary value problems for infinite systems of such equations is to start with a finite number of such equations (that is, studying finite systems of such equations and then extending the obtained results to the infinite systems). However, serious basic difficulties emerge here. The difficulty referred to above may be surmounted by considering infinite systems in the form of system (1.4) and in the diagonal form of the operators $\mathcal{F}^i$, $i \in S$. The method will consist in the construction of certain approximating sequences tending to solutions of infinite systems in suitably chosen vector spaces.

The method of successive approximation is one of the most basic and simplest methods for proving existence theorems for certain types of equations. Numerous variants of this method are well known, using different constructions of a sequence of successive approximations and different ways of proving the convergence.

To prove the existence and uniqueness of the regular (classical) and global in a time solution (i.e., the solution is defined on the whole interval $[0, \infty]$) of this Fourier first initial value problem, we apply various monotone iterative methods in vector spaces, partially ordered by positive cones. The methods of upper and lower solutions are well-known methods of proving the existence of solutions for numerous classes of initial-boundary value problems. Applying these methods requires assuming the monotonicity of the right-hand sides of the equations, i.e., the reaction functions $f^i(t, x, y, s)$, $i \in S$, with respect to the function argument y and the functional argument s. We also assume the existence of an ordered pair of a lower u_0 and an upper v_0 solutions to the problem considered. Moreover, a basic assumption on $f^i(t, x, y, s)$, $i \in S$, is also the left-hand side Lipschitz condition. These assumptions are not typical, but we obtain constructive existence theorems and *a priori* information on a sector $\langle u_0, v_0 \rangle$ in which those solutions must remain. We remark that the right-hand side Lipschitz condition is used to ensure the uniqueness of a solution.

We will assume that the reaction functions $f^i(t, x, y, s)$, $i \in S$, are Volterra functionals, i.e., they satisfy the Volterra condition with respect to the last variable s. This means that the values of these functions depend only on the past history of the process. An exceptionally important role in the theory of nonlinear equations plays the nonlinear superposition operator, whose properties are studied in certain important lemmas.

In each case, the monotone iterative techniques as the method of proving the existence of solutions consist of three steps: (i) constructing a nondecreasing and a nonincreasing sequence of lower and upper solutions; (ii) showing the uniform or almost uniform convergence of the constructed sequences of approximate solutions; and (iii) proving that the limit functions are solutions.

To examine the existence and uniqueness of a solution of infinite systems, six monotone iterative methods are employed in chapter 4. In the method of direct iteration, Chaplygin method and its modifications, the successive terms of the approximation sequences are defined as solutions of linear equations. We also present a different variant of a monotone iterative method, in which we apply the important idea of a pseudo-linearization of nonlinear problems (the successive terms of approximation sequences are defined as solutions of equations).

The main difference between the Chaplygin method and other methods of monotone iterations lies in the definitions of the approximating sequences. This method requires a stronger assumption on the functions f^i, namely the convexity assumption, but yields sequences of successive approximations converging to the solution of the problem under consideration more quickly than the iterative sequences $\{u_n\}$ and $\{v_n\}$ constructed other methods, under the obvious assumption that all monotone iterative methods start from the same pair of a lower solution u_0 and an upper solution v_0, whose existence is assumed. These sequences are converged quadratically to the searched solution.

Using Chaplygin's idea we will use the linearization of functions $f^i(t, x, y, s)$, $i \in S$, with respect to both arguments: y and s, simultaneously. A characteristic feature of this method is construction of the monotone sequences converging to the solution sought within the sector $\langle u_0, v_0 \rangle$, and whose convergent speed is quadratic. We will use this method for infinite systems of equations with functionals.

Next we present a monotone method of direct iterations in an unbounded domain, when considered functions satisfy some growth condition.

Two ideas are worth mentioning among the proposed approaches which enable us to delete the monotonicity condition for the right-hand sides with respect to the function and functional arguments (or at least the latter one). For simplicity, we will consider the case where the right-hand side depends only on the functional argument.

The first idea consists of the modification of each of the equations by adding, to both sides of the equations, the same linear term including the unknown function, to render the new right-hand sides of the equations monotone, should this be possible. This means that we will be replacing the monotonicity condition with a weaker one, namely, a semi-monotonicity condition.

The other idea consists of dividing the function arguments into two groups, depending on whether the right-hand side of an equation is increasing or decreasing with respect to a given argument. Whenever this is the case, we speak about mixed monotonicity. The splitting of the reaction function leads to four different types of ordered pairs of lower and upper solutions. The situation is best reflected in the choice of the pair of functions which start the approximation process.

It should be noted here that both ideas were developed to answer the need to study specific finite systems of equations and they admit direct extension to the infinite systems of equations.

The proofs of theorems are based on Szarski's results concerning the differential inequalities for weakly coupled infinite systems of equations in which Volterra funtionals arise. The theorems on weak and strong differential inequalities, the comparison theorem and a uniqueness criterion are given in chapter 3. The maximum principle, which plays a fundamental role (see Chapter 3) is reflected in what is known as a positivity lemma, which directly follows from the maximum principle.

The uniqueness of solutions to the problems considered in our work is guaranteed by the Lipschitz condition and follows from Szarski's uniqueness criterion.

The monotone iterative methods are quite useful for computation of numerical solutions of finite systems of equations.

Chapter 4 gives monotone iterative methods for the case of infinite systems of equations whose right-hand side depends on $z(t, \cdot)$, i.e., $f^i(t, x, z(t, x), z(t, \cdot))$.

In Chapter 5, we give some remarks on the monotone iterative methods and on the constructions of upper and lower solutions, for finite systems of equations, and some remarks concerning the possibility of extending these methods to more general infinite systems of equations.

In Chapter 6, we describe the truncation method for infinite countable and uncountable systems of equations which play a very important role among approximation methods. The application of these methods leads to sequences of approximate solutions $\{z_{N,\psi}^N\}_{N=1,2,...}$, which—under appropriate conditions—tend to the exact solution of a given problem as $N \to \infty$. Let us observe that we do not need to know the previous approximations to determine the Nth approximation as a solution of finite systems of the first N equations with N unknown functions which are called truncated systems. These finite systems of equations are discretized by the finite-difference method. There are three basic monotone iteration schemes for finite-difference systems: a Picard iteration, a modified version of a Jacobi iteration, and the Gauss-Seidel method. These iterations also provide numerical approximation of solutions which is important in practical applications. If the initial iteration is always a pair of known coupled lower and upper solutions of the considered problem, then the sequence of Picard iteration converges faster than the sequence of a Gauss-Seidel iteration which in turn converges faster than the sequence of the Jacobi iteration. These monotone iterative schemes provide the sequences of linear iterations which converge monotonically to a unique solution of the considered truncated problem. Therefore this method is very useful and commonly used in practical computation of approximate solutions.

To study the existence and uniqueness of a solution of infinite systems, in Chapter 7 we use the topological fixed point method. Considering partially ordered Banach space of continuous mappings, we give some natural sufficient conditions for the existence and uniqueness of the solution. The *a priori* estimates which appear in applications of the Banach and Schauder fixed point theorems are parallel to the above-mentioned assumptions in the theory of monotone iterative methods. Finally, to prove the existence and uniqueness of global solutions in a time, i.e., the solution defined on the whole interval $[0, \infty)$, we apply the Leray-Schauder fixed point theorem but first we extend some *a priori* estimates of the Friedman type.

In Chapter 8 the theorem on existence of a solution of the Dirichlet problem for infinite systems of equations is proved with use of a certain variant of the Chaplygin method. The results obtained are applied to study the asymptotic behavior of the global time-dependent solutions in relation to the solutions of the time-independent problem. It has been proved that the limit of the solution as $t \to \infty$ is a solution of the time-independent problem, obtained by applying the monotone method of associated lower and upper solutions.

The tools of modern mathematics in the development of a framework that underpins the understanding of infinite systems in general is developed (Part I). Infinite systems of equations arise in a variety of physical contexts, especially when modeling in the neurosciences. New insights can be achieved that will lead to a better understanding of the brain and nervous system from a mathematical perspective (see Part II).

CHAPTER 2

Preliminary Considerations

2.1 Sets and Domains

We shall say that a set D (possibly unbounded) in the time-space $(t, x) = (t, x_1, x_2, \ldots, x_m)$ has *property* $\mathcal{P}$ if:

i.   the projection of the interior of D on the t-axis is the interval $(0, T)$, where $0 < T < \infty$;
ii. for every $(\tilde{t}, \tilde{x}) \in D$, there exists a positive number r such that the left neighborhood is contained in D, i.e.,

$$\{(t, x) \colon (t - \tilde{t})^2 + |x - \tilde{x}|^2 < r^2, \ t \le \tilde{t}\} \subset D,$$

where $|x| := \left(\sum_{j=1}^{m} x_j^2 \right)^{\frac{1}{2}}$.

Let D be a set having property $\mathcal{P}$. By D we denote an arbitrary fixed set which fulfills the following conditions:

$$D_0 \subset \{(t, x) \colon t \le T, \ x \subset \mathbb{R}^m\} \quad \text{and} \quad \overline{D} \subset D.$$

We denote by σ the part of the boundary ∂D of domain D situated in the open zone $\{(t, x) \colon 0 < t < T, \ T < \infty, x \in \mathbb{R}^m\}$ (this is the lateral surface of domain D). For an arbitrary fixed $\tau, 0 \le \tau \le T$, we define

$$S_\tau := D \cap \{(t, x) \colon t = \tau, \ x \in \mathbb{R}^m\} = \{(t, x) \in D \colon t = \tau\}$$

and we assume that $S_0 := \{(t, x) \in D \colon t = 0, \ x \in \mathbb{R}^m\}$ is a nonempty set. We denote by $\Gamma := S_0 \cup \sigma$ the parabolic boundary of domain D and $\overline{D} := D \cup \Gamma$.

2.2 Banach and Hölder Spaces

Let D be a domain in the time-space $(t, x) = (t, x_1, x_2, \ldots, x_m)$ and S be an arbitrary set of indices (finite or infinite).

Mathematical Neuroscience. http://dx.doi.org/10.1016/B978-0-12-411468-5.00002-8

Let $\mathcal{B}(S)$ be the real space of mappings

$$w: S \to \mathbb{R}, \quad i \mapsto w(i) := w^i,$$

such that

$$\sup\{|w^i| : i \in S\} < \infty$$

equipped with the supremum norm

$$\|w\|_{\mathcal{B}(S)} := \sup\{|w^i| : i \in S\}.$$

This space is a Banach space.

The space ℓ^∞ is the sequence space of all real-valued bounded sequences $w = \{w^j\}_{j \in \mathbb{N}} = (w^1, w^2, \dots)$ such that

$$\sup\{|w^j| : j \in \mathbb{N}\} < \infty$$

equipped with the norm

$$\|w\|_{\ell^\infty} := \sup\{|w^j| : j \in \mathbb{N}\}.$$

If S is a finite set of indices with r elements, i.e., $S = \{1, 2, \dots, r\}$ then $\mathcal{B}(S) := \mathbb{R}^r$ and for an infinite countable set S, there is $\mathcal{B}(S) := \mathcal{B}(\mathbb{N}) = \ell^\infty$.

Now, we introduce the spaces of sequences of real-valued functions equipped with the norms induced by the norm of the space ℓ^∞.

Denote by $\mathscr{C}_\mathbb{N}(\overline{D}) := \mathscr{C}_\mathbb{N}^0(\overline{D})$ the space of infinite sequences $w = (w^1, w^2, \dots)$ of real-valued functions $w^j = w^j(t, x)$, $j \in \mathbb{N}$, defined and continuous in a domain $\overline{D}$, such that

$$\sup\{|w^j| : j \in \mathbb{N}\} < \infty$$

equipped with the norm

$$\|w\|_{\mathscr{C}_\mathbb{N}(\overline{D})} := \sup\{|w^j|_0 : j \in \mathbb{N}\},$$

where $w^j \in C(\overline{D})$, $j \in \mathbb{N}$, and

$$|w^j|_0 := \sup\{|w^j(t, x)| : (t, x) \in \overline{D}\} < \infty$$

is the norm in the space $C(\overline{D})$ of all functions continuous in a domain $\overline{D}$.

We introduce the space $\mathscr{C}_{N,0}(\overline{D})$, consisting of those infinite sequences in $\mathscr{C}_\mathbb{N}(\overline{D})$ which have the following form $w_{N,0} = (w_N^1, w_N^2, \dots, w_N^N, 0, 0, \dots)$ where $w_N^j \in \mathbb{R}$ for $j = 1, 2, \dots, N$ and $w_N^j \equiv 0$ for $j = N + 1, N + 2, \dots$ such that

$$\max\left\{\left|w_N^j\right|_0 : j = 1, 2, \dots, N\right\} < \infty$$

equipped with the norm

$$\left\| w_{N,0} \right\|_{\mathscr{C}_{N,0}(\overline{D})} = \max \left\{ \left| w_N^j \right|_0 : j = 1, 2, \ldots, N \right\}.$$

A real function $h = h(x)$ defined on a bounded closed set $A \subset \mathbb{R}^m$ is said to be *Hölder continuous with exponent* α $(0 < \alpha \le 1)$ in A if there exists a constant $H > 0$ such that

$$\left| h(x) - h(x') \right| \le H \left| x - x' \right|^\alpha \quad \text{for all } x, x' \in A.$$

The smallest H for which this inequality holds is called the *Hölder coefficient*.

If $\alpha = 1$, then we say that $h = h(x)$ is *Lipschitz continuous* in A.

If A is an open set then h is *locally Hölder continuous* with exponent α in A; if this inequality holds in every bounded closed subset $B \subset A$, where H may depend on B. If H is independent of B, then we say that h is *uniformly Hölder continuous* in A.

If the function h also depends on a parameter λ, i.e., $h = h(x, \lambda)$, and if the Hölder coefficient H is independent of λ, then we say that h is Hölder continuous in x, uniformly with respect to λ.

The real function $w = w(t, x)$ defined on a bounded closed set $D \subset \mathbb{R}^{m+1}$ is said to be Hölder continuous with respect to t and x with exponent α $(0 < \alpha < 1)$ in $\overline{D}$ if there exists a constant $H > 0$ such that

$$|w(t, x) - w(t', x')| \le H(|t - t'|^{\frac{\alpha}{2}} + |x - x'|^\alpha)$$

for all $(t, x), (t', x') \in \overline{D}$.

The *Hölder space* $C^{k+\alpha}(\overline{D})$ $(k = 0, 1, 2; 0 < \alpha < 1)$ is a Banach space of functions $w = w(t, x)$, that are continuous in $\overline{D}$, together with all derivatives of the form $\mathcal{D}_t^r \mathcal{D}_x^s w(t, x)$ for $0 \le 2r + s \le k$, and have the finite *norm*

$$|w|_{k+\alpha} := \sup_{\substack{P \in D \\ 0 \le 2r+s \le k}} \left| \mathcal{D}_t^r \mathcal{D}_x^s w(P) \right| + \sup_{\substack{P, P' \in D \\ 2r+s=k \\ P \ne P'}} \frac{\left| \mathcal{D}_t^r \mathcal{D}_x^s w(P) - \mathcal{D}_t^r \mathcal{D}_x^s w(P') \right|}{[d(P, P')]^\alpha},$$

where $d(P, P') = (|t - t'| + |x - x'|^2)^{\frac{1}{2}}$ is the *parabolic distance of points* $P(t, x)$, $P'(t', x') \in \mathbb{R}^{m+1}$.

In particular, there is

$$|w|_{0+\alpha} = |w|_0 + H_\alpha^D(w), \quad |w|_0 = \sup_{P \in D} |w(P)|, \quad H_\alpha^D(w) = \sup_{\substack{P, P' \in D \\ P \ne P'}} \frac{|w(P) - w(P')|}{[d(P, P)]^a},$$

$$|w|_{1+\alpha} = |w|_{0+\alpha} + \sum_{j=1}^{m} |\mathcal{D}_{x_j} w|_{0+\alpha},$$

$$|w|_{2+\alpha} = |w|_{0+\alpha} + \sum_{j=1}^{m} |\mathcal{D}_{x_j} w|_{0+\alpha} + \sum_{j,k=1}^{m} |\mathcal{D}^2_{x_j x_k} w|_{0+\alpha} + |\mathcal{D}_t w|_{0+\alpha}.$$

$H_\alpha^D(w) < \infty$ if and only if w is uniformly Hölder continuous (with exponent α) in D, and then $H_\alpha^D(w)$ is the Hölder coefficient of w.

By $C_{k+\alpha,S}(\overline{D}) := C_S^{k+\alpha}(\overline{D})$ ($k = 0, 1, 2$; $0 < \alpha < 1$) we denote a Banach *space* of functions $w = \{w^i\}_{i \in S}$ such that $w^i \in C^{k+\alpha}(\overline{D})$ for all $i \in S$, with the finite *norm*

$$\|w\|_{k+\alpha} := \sup\left\{|w^i|_{k+\alpha} : i \in S\right\}.$$

Remark 2.1 If S is a finite set with r elements then we will denote suitable spaces by $C_r^{k+\alpha}(\overline{D})$.

The *boundary norm* $\|\cdot\|_{k+\alpha}^\Gamma$ ($k = 0, 1, 2$; $0 < \alpha < 1$) of a function $\phi \in C_S^{k+\alpha}(\Gamma)$ is defined as

$$\|\phi\|_{k+\alpha}^\Gamma := \inf_\Phi \|\Phi\|_{k+\alpha},$$

where the infimum is taken over the set of all possible extensions Φ of ϕ onto $\overline{D}$ such that $\Phi \in C_S^{k+\alpha}(\overline{D})$.

We denote by $C^{k-0}(D)$ ($k = 1, 2$) the *spaces* of functions $w \in C(\overline{D})$ for which the following norms are finite

$$|w|_{1-0} := |w|_0 + \sup_{\substack{P, P' \in D \\ P \neq P'}} \frac{|w(t, x) - w(t', x')|}{|t - t'| + |x - x'|},$$

$$|w|_{2-0} := |w|_{1-0} + \sum_{j=1}^{m} |w_{x_j}|_{1-0}.$$

2.3 Cones and Ordered Spaces

In this book, we will consider Banach spaces with a partial order induced by a cone.

Let $\mathcal{X}$ be a real Banach space. A proper subset K of $\mathcal{X}$ is said to be a *cone* if $\lambda K \subset K$ for every $\lambda \geq 0$, $K + K \subset K$, $K \cap (-K) = \{0\}$, where 0 denotes the null element of the Banach space $\mathcal{X}$ and $K = \overline{K}$, where $\overline{K}$ denotes the closure of K.

We will consider a *closed cone* K and its interior $K^0 := int\, K = Q$, which is an *open cone*.

The *partial order* "$\leq$" in the Banach space $\mathcal{X}$ with the closed cone K may be defined in the following way:

$$u \leq v \Longleftrightarrow v - u \in K,$$

and then $\mathcal{X}$ is called *a partially ordered Banach space.*

The open cone Q such that $Q = K^0$ induces the order relation "$<$" in the Banach space $\mathcal{X}$ defined by

$$u < v \Longleftrightarrow v - u \in Q$$

and the elements of Q are called positive.

The partial order "$\leq$" in the space l^∞ is defined by the *positive (nonnegative) cone*

$$l_+^\infty := \left\{ w : w = \{w^i\}_{i \in \mathbb{N}} \in l^\infty, \ w^i \geq 0 \text{ for } i \in \mathbb{N} \right\}$$

in the following way:

$$u \leq v \Longleftrightarrow v - u \in l_+^\infty.$$

Analogously, the partial order in the space $\mathcal{B}(S)$ is defined by the positive (nonnegative) cone

$$\mathcal{B}^+(S) := \left\{ w : w = \{w^i\}_{i \in S} \in \mathcal{B}(S), \ w^i \geq 0 \text{ for } i \in S \right\}$$

in the following way:

$$u \leq v \Longleftrightarrow v - u \in \mathcal{B}^+(S).$$

The partial order "$\leq$" in the space $C_S(\overline{D})$ is defined by means of the positive closed cone

$$K := C_S^+(\overline{D})$$
$$:= \left\{ w : w = \{w^i\}_{i \in S} \in C_S(\overline{D}), \ w^i(t, x) \geq 0 \text{ for } (t, x) \in \overline{D} \text{ and } i \in S \right\}$$

in the following way:

$$u \leq v \Longleftrightarrow v - u \in C_S^+(\overline{D}).$$

The partial order in the Hölder space $C_S^{k+\alpha}(\overline{D}) := C_{k+\alpha, S}(\overline{D}), \ (k = 0, 1, 2; 0 < \alpha < 1)$ is defined by means of the positive cone

$$C_{k+\alpha, S}^+(\overline{D}) := \left\{ w : w = \{w^i\}_{i \in S} \in C_{k+\alpha, S}(\overline{D}), \right.$$

$$\left. w^i(t, x) \geq 0 \ \text{ for } (t, x) \in \overline{D} \text{ and } i \in S \right\}$$

in the following way:

$$u \leq v \Longleftrightarrow v - u \in C_{k+\alpha, S}^+(\overline{D}).$$

Remark 2.2 From this it follows that the inequality $u(t, x) \le v(t, x)$ is to be understood componentwise, i.e., $u^i(t, x) \le v^i(t, x)$ for all $i \in S$.

Inequality $u \le v$ is to be understood both componentwise and pointwise, i.e., $u^i(t, x) \le v^i(t, x)$ for arbitrary $(t, x) \in D$ and all $i \in S$.

The notation $u < v$ means $u^i(t, x) < v^i(t, x)$ for arbitrary $(t, x) \in D$ and all $i \in S$.[1]

2.4 Ellipticity and Parabolicity

In the book we will consider the following operators:

$$\mathcal{F}_a^i := \mathcal{D}_t - \mathcal{A}^i, \quad \mathcal{A}^i := \sum_{j,k=1}^m a_{jk}^i(t, x) \mathcal{D}_{x_j x_k}^2,$$

$$\mathcal{F}^i := \mathcal{D}_t - \mathcal{L}^i, \quad \mathcal{L}^i := \sum_{j,k=1}^m a_{jk}^i(t, x) \mathcal{D}_{x_j x_k}^2 - \sum_{j=1}^m b_j^i(t, x) \mathcal{D}_{x_j},$$

and

$$\mathcal{F}_c^i := \mathcal{D}_t - \mathcal{L}_c^i, \quad \mathcal{L}_c^i := \sum_{j,k=1}^m a_{jk}^i(t, x) \mathcal{D}_{x_j x_k}^2 - \sum_{j=1}^m b_j^i(t, x) \mathcal{D}_{x_j} - c^i(t, x) \mathcal{I}, \quad i \in S,$$

where $a_{jk}^i(t, x) = a_{ki}^j(t, x)$, for $(t, x) \in D$, $k = 1, 2, \ldots, m$ and $i \in S$ where $\mathcal{I}$ is the identity operator.

We will assume that all coefficients of these operators are continuous in domain D or in their respective domains.

Let us consider the equation of the form

$$\mathcal{F}_a[z](t, x) := \mathcal{D}_t z(t, x) - \mathcal{A}[z](t, x) = f(t, x, z(t, x), z)$$

in domain $D \subset \mathbb{R}^{m+1}$, where

$$\mathcal{A} := \sum_{j,k=1}^m a_{jk}(t, x) \mathcal{D}_{x_j x_k}^2, \quad a_{jk}(t, x) = a_{kj}(t, x) \quad \text{for } (t, x) \in D \quad \text{and} \quad j, k = 1, 2, \ldots, m.$$

---

[1] The partial ordering in a set $\mathcal{X}$ also induces a corresponding partial ordering in a subset W of $\mathcal{X}$ and if $u, v \in W$ with $u \le v$, then

$$\langle u, v \rangle := \{s \in W, u \le s \le v\}$$

denotes the *sector* (*conical segment*, *order interval*), formed by the ordered pair u and v.

The operator $\mathcal{A}$ is called *elliptic at a point* $(t^*, x^*) \in D$ if and only if the coefficient matrix $(a_{jk}(t^*, x^*))$ is positive defined, i.e., if

$$\sum_{j,k=1}^{m} a_{jk}(t^*, x^*)\xi_j\xi_k > 0 \quad \text{for all } \xi \in \mathbb{R}^m \setminus \{0\},$$

where $\xi = (\xi_1, \xi_2, \ldots, \xi_m)$ here and below.

The operator $\mathcal{A}$ is said to be *elliptic in the domain* D if it is elliptic at every point of D.

The operator $\mathcal{A}$ is called *uniformly elliptic* in $\overline{D}$ if there exists a positive constant μ_0 independent of $(t, x) \in \overline{D}$ and ξ, such that the following inequality holds:

$$\sum_{j,k=1}^{m} a_{jk}(t, x)\xi_j\xi_k \geq \mu_0|\xi|^2 \quad \text{for all } \xi \in \mathbb{R}^m \quad \text{and} \quad (t, x) \in \overline{D},$$

where $|\xi|^2 = \sum_{j=1}^{m} \xi_j^2$.

If the operator $\mathcal{A}$ is elliptic and the coefficients a_{jk} are continuous in a bounded and closed domain, then $\mathcal{A}$ is uniformly elliptic in this domain.

If the *operator $\mathcal{A}$ is elliptic* (*uniformly elliptic*) in D, then the *operator $\mathcal{F}_a$ is called parabolic* (*uniformly parabolic*) in D and the related *equation is called parabolic in D*.

Let us consider the infinite system of equations of the form

$$\mathcal{F}_a^i[z^i](t, x) := \mathcal{D}_t z^i(t, x) - \mathcal{A}^i[z^i](t, x) = f^i(t, x, z(t, x), z) \tag{2.1}$$

for $(t, x) \in D$ and $i \in S$, where

$$\mathcal{A}^i := \sum_{j,k=1}^{m} a_{jk}^i(t, x)\mathcal{D}_{x_jx_k}^2, \quad a_{jk}^i(t, x) = a_{kj}^i(t, x) \quad \text{for } j, k = 1, 2, \ldots, m, \quad \text{and} \quad i \in S,$$

and S is an arbitrary set of indices (finite or infinite).

The operators $\mathcal{A}^i$, $i \in S$, are *uniformly elliptic in $\overline{D}$* if there exists a constant $\mu_0 > 0$ independent of $(t, x) \in \overline{D}$, ξ and i, such that the following inequalities hold:

$$\sum_{j,k=1}^{m} a_{jk}^i(t, x)\xi_j\xi_k \geq \mu_0|\xi|^2 \quad \text{for all } \xi \in \mathbb{R}^m, \quad (t, x) \in \overline{D}, \quad i \in S.$$

If the operators $\mathcal{A}^i$, $i \in S$, are uniformly elliptic in $\overline{D}$ then the *operators $\mathcal{F}_a^i$, $i \in S$, are uniformly parabolic* in $\overline{D}$ and the system (2.1) is called uniformly parabolic in $\overline{D}$.

2.5 Notations of Functional Dependence

Let us consider a system of equations of the form

$$\mathcal{F}^i[z^i](t, x) = f^i(t, x, z(t, x), z) \quad \text{for } (t, x) \in D, \quad i \in S, \tag{2.2}$$

where

$$f^i : \overline{D} \times \mathcal{B}(S) \times C_S(\overline{D}) \to \mathbb{R}, \quad (t, x, y, s) \mapsto f^i(t, x, y, s), \quad i \in S.$$

If we define the function $\tilde{f} = \{\tilde{f}^i\}_{i \in S}$ setting

$$\tilde{f}^i(t, x, z) := f^i(t, x, z(t, x), z), \quad i \in S,$$

where

$$\tilde{f} : \overline{D} \times C_S(\overline{D}) \to \mathbb{R}, \quad (t, x, s) \mapsto \tilde{f}^i(t, x, s), \quad i \in S,$$

then we will write system (2.2) in another form

$$\mathcal{F}^i[z^i](t, x) = \tilde{f}^i(t, x, z), \quad i \in S, \quad \text{for } (t, x) \in D, \tag{2.3}$$

which may be useful in further considerations.

Both forms (2.2) and (2.3) of the system considered have some disadvantages. Namely, if the right-hand sides of equations containing operators are of the form

$$\tilde{f}(t, x, z) \quad \text{or} \quad f(t, x, z(t, x), z),$$

then it is difficult to formulate conditions concerning the existence of solutions.

The situation is different if the right-hand sides are of the form

$$\hat{f}^i(t, x, z(t, x), V[z](t, x)), \quad i \in S,$$

i.e., are superpositions of functions defined on certain subsets of the space $\mathbb{R}^{m+1} \times \mathcal{B}(S) \times \mathcal{B}(S)$ and some operators $V = \{V^i\}_{i \in S}$.

Thus

$$\hat{f}^i : \overline{D} \times \mathcal{B}(S) \times \mathcal{B}(S) \to \mathbb{R}, \quad (t, x, y, s) \mapsto \hat{f}^i(t, x, y, s), \quad i \in S$$

and

$$V^i : C_S(\overline{D}, \mathbb{R}) \to \mathbb{R}, \quad z \mapsto V^i[z], \quad i \in S,$$

that is

$$V : C_S(\overline{D}, \mathbb{R}) \to \mathcal{B}(S), \quad z \mapsto V[z].$$

Therefore, system (2.3) will take the following form:

$$\mathcal{F}^i[z^i](t, x) = \hat{f}^i(t, x, z(t, x), V[z](t, x)), \quad i \in S, \tag{2.4}$$

where $V[z] = \{V^i[z]\}_{i \in S}$.

Now we may adopt separate assumptions on the functions $\hat{f}^i$ and operator V. We may assume different conditions concerning the type of dependence of $\hat{f}^i$ on the variables t and x.

The objective of using different expressions is to facilitate the understanding of the problem, as well as to facilitate, if not make possible, a precise formulation and wording of appropriate assumptions.

Finally, it should be noted that, formally, all three expressions (2.3) and (2.4) are equivalent and may be used alternately, as the context requires.

2.6 Initial and Boundary Conditions

Now, for system (2.2), we will consider the Fourier first problem with the initial-boundary condition

$$z(t, x) = \phi(t, x) \quad \text{for } (t, x) \in \Gamma, \tag{2.5}$$

or with the *homogeneous initial-boundary condition*

$$z(t, x) = 0 \quad \text{for } (t, x) \in \Gamma, \tag{2.6}$$

where Γ is the parabolic boundary of domain D.

If we consider a cylindrical domain $D := (0, T] \times G$, where T is an arbitrary positive constant and G is an open and bounded domain in $\mathbb{R}^m$, then (here and below) we will use notations: $S_0 := \{(t, x) : t = 0, \ x \in \overline{G}\}$, the lateral surface $\sigma := (0, T] \times \partial G$, the parabolic boundary of domain D is $\Gamma := S_0 \cup \sigma$, and $\overline{D} := D \cup \Gamma = [0, T] \times \overline{G}$.

Most frequently we will write condition in another form, i.e., separately as the Cauchy initial condition

$$z(0, x) = \phi_0(x) \quad \text{for } x \in G \tag{2.7}$$

and as the *Dirichlet boundary condition*

$$z(t, x) = h(t, x) \quad \text{for } (t, x) \in \sigma \tag{2.8}$$

with the suitable *compatibility condition*

$$h(0, x) = \phi_0(x) \quad \text{for } x \in \partial G. \tag{2.9}$$

The compatibility conditions for functions f, ϕ_0, and h which appear in (2.2), (2.7), and (2.8) state that the derivative $\mathcal{D}_t z^i \big|_{t=0}$, $i \in S$, which is determined for $t = 0$ by means of the equations of the system and of the initial condition, also satisfies the boundary condition for $x \in \partial G$. Therefore, in our case these functions must satisfy the *compatibility conditions of order* $\left[\frac{\alpha}{2}\right] + 1$ of the form

$$h(0, x) = \phi_0(x), \quad \mathcal{F}^i[h^i](0, x) = f^i(0, x, h(0, x), h), \quad i \in S, \quad \text{for } x \in \partial G. \tag{2.10}$$

For system (2.2), we may consider the *Neumann boundary condition*

$$\frac{dz(t, x)}{dv} = g(t, x) \quad \text{for } (t, x) \in \sigma \tag{2.11}$$

or the *Robin boundary condition*

$$\frac{dz(t, x)}{dv} + \beta(t, x)z(t, x) = g(t, x) \quad \text{for } (t, x) \in \sigma, \tag{2.12}$$

where $\frac{d}{dv}$ is the directional derivative in the direction v on σ and in this book, $\frac{dz}{dv}$ is considered as the outward normal derivative of z on σ with the suitable zero-order compatibility conditions.

A particular case of (2.11) is the sealed-end boundary condition

$$\frac{dz(t, x)}{dv} = 0 \quad \text{for } (t, x) \in \sigma$$

corresponding to the assumption of the domain G being ideally isolated from the surrounding medium so that no charge crosses its boundary.

We define the boundary operator $B = \{B^i\}_{i \in S}$ by setting

$$B^i[z^i](t, x) := \alpha^i(t, x)\frac{dz^i(t, x)}{dv} + \beta^i(t, x), \quad i \in S, \quad \text{for } (t, x) \in \sigma, \tag{2.13}$$

where the coefficients $\alpha = \{\alpha^i\}_{i \in S}$, $\alpha^i = \alpha^i(t, x)$, and $\beta = \{\beta^i\}_{i \in S}$, $\beta^i = \beta^i(t, x)$ are nonnegative continuous functions defined for $(t, x) \in \sigma$ and $\alpha(t, x) + \beta(t, x) > 0$.

If we consider the boundary condition of the form

$$B[z](t, x) = g(t, x) \quad \text{for } (t, x) \in \sigma, \tag{2.14}$$

then the latter assumption includes the Dirichlet condition ($\alpha = 0$, $\beta = 1$), Neumann condition ($\alpha = 1$, $\beta = 0$), and Robin condition ($\alpha = 1$, $\beta \geq 0$).

These are, in particular, nonlinear nonlocal boundary conditions of the form

$$B[z](t, x) = g(t, x, z(t, x), z) \quad \text{for } (t, x) \in \sigma. \tag{2.15}$$

When we say that a boundary condition is nonlocal, we mean that the function g may depend on the unknown function z, not only functionwise, but also functionalwise.

The simplest possible formulation of a nonlinear, nonlocal boundary condition and the comparison thereof with the classical Dirichlet boundary condition is illustrated for the Poisson equation

$$\Delta z = f(x) \quad \text{in } G, \tag{2.16}$$

where $G \subset \mathbb{R}^3$ is a bounded domain with the boundary ∂G. Let Γ_1 and Γ_2 be open subsets of ∂G and let

$$\Gamma_1 \cap \Gamma_2 = \emptyset, \quad \partial \Gamma_1 = \partial \Gamma_2 = \Gamma_0, \quad \text{and} \quad \Gamma_1 \cup \Gamma_2 \cup \Gamma_0 = \partial G.$$

We introduce a mapping

$$F : \overline{\Gamma}_2 \times C(\overline{G}) \to \mathbb{R}.$$

We assume that the functions f, φ, and ψ are defined on G, Γ_1, and Γ_2, respectively. For (2.16), we will consider the following boundary value problem:

Find a solution $z \in C^2(G) \cap C(\overline{G})$ of (2.16) satisfying the mixed boundary condition

$$z(x) = \varphi(x) \quad \text{for } x \in \overline{\Gamma}_1 \tag{2.17}$$

and

$$z(x) + F(x, z(\cdot)) = \psi(x) \quad \text{for } x \in \overline{\Gamma}_2. \tag{2.18}$$

Condition (2.18) determines the nonlocal nature of problem (2.16)–(2.18) in which the values of a solution on the part Γ_2 of the boundary ∂G are connected with values of z on G.

The existence and uniqueness of the solution of the Cauchy problem for infinite systems of equations with Volterra functionals of the form

$$\begin{cases} \mathcal{F}^i[z^i](t, x) = f^i(t, x, z(t, x), z) & \text{for } (t, x) \in \Omega := (0, T) \times \mathbb{R}^m, \\ z(0, x) = \phi_0(x) & \text{for } x \in \mathbb{R}^m, \quad i \in S, \end{cases} \tag{2.19}$$

where $\Omega := (0, T) \times \mathbb{R}^m$.

This problem can be studied with use of the monotone method of direct iterations and the topological fixed point method (the Banach contraction principle and the Schauder fixed point theorem) in the class of continuous and bounded functions and in the class of functions satisfying a certain growth condition. The existence and uniqueness of a bounded solution of this problem under Carathéodory-type conditions can be proved by using the Banach contraction principle.

Such problems frequently appear in applications, when it is very difficult or just impossible to describe a boundary of the domain or boundary condition under consideration.

Below we present a couple of interesting examples of nonlocal problems:

$$\begin{cases} \dfrac{\partial u}{\partial t} - \mathcal{L}[u] = f(t, x, u) & \text{for } (t, x) \in D := (0, T] \times G, \\ \mathcal{B}[u] = \displaystyle\int_G K(x, y)\, u(t, y)\, dy & \text{for } (t, x) \in \sigma, \\ u(0, x) = u_0(x) & \text{for } x \in G. \end{cases}$$

$$
\begin{cases}
\dfrac{\partial u}{\partial t} - \mathcal{L}[u] = f(t, x, u) & \text{for } (t, x) \in D := (0, T] \times G, \\[3mm]
\mathcal{B}[u] = \displaystyle\int_G K(x, y)\, u(t, y)\, dy + h(x) & \text{for } (t, x) \in \sigma, \\[3mm]
u(0, x) = \displaystyle\int_0^\infty g(t, x)\, u(t, x)\, dt + \Psi(x) & \text{for } x \in G.
\end{cases}
$$

$$
\begin{cases}
\dfrac{\partial u^i}{\partial t} - \mathcal{L}^i[u^i] = f^i(t, x, u) & \text{for } (t, x) \in D, \\[3mm]
\dfrac{\partial u^i}{\partial \nu_i} = g^i(t, x, u(\cdot, \cdot)) \quad \text{or} \quad u(t, x) = 0 & \text{for } (t, x) \in \sigma, \\[3mm]
u(t, x) = \psi(t, x) & \text{for } (t, x) \in (-r, 0) \times G,
\end{cases}
$$

where $u = (u^1, u^2, \ldots, u^r)$, $i = 1, 2, \ldots, r$, and the nonlinearities f and g may contain functionals of u such as

$$
\int_0^t K(t, s)\, u(s, x)\, ds, \qquad \int_{|\xi| \le |x|^2} u(t, x + \xi)\, d\xi,
$$

$$
|u(t - \tau(t, x),\ p(t, x))|^\lambda,
$$

$$
\int_G \left| \frac{\partial u}{\partial x}(t - \tau(t, x), \xi) \right|^2 d\xi, \quad \text{where } \tau(t, x) \ge 0 \ \text{ and } \ \lambda > 0.
$$

2.7 Fundamental Assumptions and Conditions

Now we will introduce the following assumptions on the coefficients of considered operators $\mathcal{A}^i$, $\mathcal{L}^i$, and $\mathcal{L}_c^i$ and f^i for $i \in S$ as follows:

Assumption H_a. We will assume that all the coefficients $a_{jk}^i = a_{jk}^i(t, x)$, $a_{jk}^i = a_{kj}^i$, and $b_j^i = b_j^i(t, x)$ $(j, k = 1, \ldots, m,\ i \in S)$ of the operators $\mathcal{L}^i$, $i \in S$, are uniformly Hölder continuous with respect to t and x in $\overline{D}$ with exponent α $(0 < \alpha < 1)$ and their Hölder norms are uniformly bounded, i.e.,

$$
\left| a_{jk}^i \right|_{0+\alpha} \le k, \quad \left| b_j^i \right|_{0+\alpha} \le k \quad \text{for } j, k = 1, \ldots, m, \quad i \in S,
$$

where k is a positive constant.

Assumption H_c. We will assume that all the coefficients $a_{jk}^i = a_{jk}^i(t, x)$, $a_{jk}^i = a_{kj}^i$, $b_j^i = b_j^i(t, x)$, and $c^i = c^i(t, x)$ $(j, k = 1, \ldots, m, i \in S)$ for the operators $\mathcal{L}_c^i, i \in S$, satisfy condition (H_a) and

$$
|c^i|_{0+\alpha} \le k \quad \text{for } i \in S.
$$

Assumption H_a^*. We will assume that the coefficients $a_{jk}^i = a_{jk}^i(t, x)$, $a_{jk}^i = a_{kj}^i$ $(j, k = 1, \ldots, m, \ i \in S)$ of the operators $\mathcal{A}^i$, $i \in S$, are uniformly Hölder continuous with exponent α $(0 < \alpha < 1)$ in $\overline{D}$, $a_{jk}^i \in C^{0+\alpha}(\overline{D})$ and a_{jk}^i belong to $C^{1-0}(\sigma)$.

From this there follows the existence of constants $k_1, k_2 > 0$ such that

$$\sum_{j,k=1}^{m} \left| a_{jk}^i \right|_{0+\alpha} \le k_1, \quad \sum_{j,k=1}^{m} \left| a_{jk}^i \right|_{1-0}^{\Gamma} \le k_2, \quad i \in S.$$

We will consider the following infinite systems of equations:

$$\mathcal{F}^i[z^i](t, x) = f^i(t, x, z(t, x), z), \quad i \in S$$

and

$$\mathcal{F}^i[z^i](t, x) = \tilde{f}^i(t, x, z), \quad i \in S,$$

where

$$f^i : \overline{D} \times \mathcal{B}(S) \times C_S(\overline{D}) \to \mathbb{R}, \quad (t, x, y, s) \mapsto f^i(t, x, y, s), \quad i \in S$$

and

$$\tilde{f}^i : \overline{D} \times C_S(\overline{D}) \to \mathbb{R}, \quad (t, x, s) \mapsto \tilde{f}^i(t, x, s), \quad i \in S.$$

We will assume that the nonlinear reaction functions $f^i(t, x, y, s)$ or $f^l(t, x, s)$, $i \in S$, are continuous functions in respective domains and satisfy the suitable assumptions.

We will introduce the following assumptions and conditions:

Assumption H_f. Functions $f^i(t, x, y, s)$, $i \subset S$, are *uniformly Hölder continuous* with exponent α $(0 < \alpha < 1)$ with respect to t and x in $\overline{D}$, and their Hölder norms $|f^i|_{0+\alpha}$ are uniformly bounded, i.e., $f(\cdot, \cdot, s) \in C_S^{0+\alpha}(\overline{D})$.

Condition L. Functions $f^i(t, x, y, s)$, $i \in S$, fulfill the *Lipschitz condition* with respect to y and s, if for arbitrary $y, \tilde{y} \in \mathcal{B}(S)$ and $s, \tilde{s} \in C_S(\overline{D})$ the inequality

$$\left| f^i(t, x, y, s) - f^i(t, x, \tilde{y}, \tilde{s}) \right| \le L_1 \|y - \tilde{y}\|_{\mathcal{B}(S)} + L_2 \|s - \tilde{s}\|_0 \quad \text{for } (t, x) \in \overline{D}$$

holds, where L_1 and L_2 are positive constants.

Condition L_l. Functions $f^i(t, x, y, s)$, $i \in S$, fulfill the *left-hand side Lipschitz condition* with respect to y, if for arbitrary $y, \tilde{y} \in \mathcal{B}(S)$ the inequality

$$-L_1 \|y - \tilde{y}\|_{\mathcal{B}(S)} \le f^i(t, x, y, s) - f^i(t, x, \tilde{y}, s) \quad \text{for } (t, x) \in \overline{D}, \quad s \in C_S(\overline{D})$$

holds, where L_1 is a positive constant.

Condition L_l^i. Function $f^i(t, x, y, s)$ satisfies for every fixed $i \in S$ the *left-hand side generalized Lipschitz condition* with respect to y^i, and if there exist bounded functions $\underline{l}^i = \underline{l}^i(t, x) \geq 0$ in $\overline{D}$ such that for $y, \tilde{y} \in B(S)$, $y \leq_{(i)} \tilde{y}$, the following inequality:

$$-\underline{l}^i(t, x)(\tilde{y}^i - y^i) \leq f^i(t, x, \tilde{y}, s) - f^i(t, x, y, s) \quad \text{for } (t, x) \in \overline{D}, \quad s \in C_S(\overline{D})$$

holds.

Monotonicity conditions.

Condition W. Functions $f^i(t, x, y, s)$, $i \in S$, are *increasing*[2] with respect to the functional argument s, i.e., for arbitrary $s, \tilde{s} \in C_S(\overline{D})$, there is

$$s \leq \tilde{s} \implies f^i(t, x, y, s) \leq f^i(t, x, y, \tilde{s}) \quad \text{for } (t, x) \in \overline{D}, \quad y \in B(S).$$

Condition W_+. Functions $f^i(t, x, y, s)$, $i \in S$, are *semi-increasing* with respect to y (or satisfy condition (W_+)), if for arbitrary $y, \tilde{y} \in B(S)$ there is

$$y \overset{(i)}{\leq} \tilde{y} \implies f^i(t, x, y, s) \leq f^i(t, x, \tilde{y}, s) \quad \text{for } (t, x) \in \overline{D}, \quad s \in C_S(\overline{D}).$$

Condition $\tilde{W}$. Functions $f^i(t, x, y, s)$, $i \in S$, are increasing with respect to y and s.

Condition V. Functions $f^i(t, x, y, s)$, $i \in S$, are *Volterra functionals* (or satisfy the *Volterra condition* (V)), if for arbitrary $(t, x) \in \overline{D}$, $y \in B(S)$ and for all $s, \tilde{s} \in C_S(\overline{D})$ such that $s^j(\bar{t}, x) = \tilde{s}^j(\bar{t}, x)$ for $0 \leq \bar{t} \leq t$, $j \in S$, there is $f^i(t, x, y, s) = f^i(t, x, y, \tilde{s})$, or shortly

$$s \overset{(t)}{=} \tilde{s} \implies f^i(t, x, y, s) = f^i(t, x, y, \tilde{s}).$$

Condition L^*. Functions $f^i(t, x, y, s)$, $i \in S$, fulfill the *Lipschitz-Volterra condition* with respect to the functional argument s if for arbitrary $s, \tilde{s} \in C_S(\overline{D})$ the inequality

$$\left| f^i(t, x, y, s) - f^i(t, x, y, \tilde{s}) \right| \leq L \|s - \tilde{s}\|_{0,t} \quad \text{for } (t, x) \in \overline{D}, \quad y \in B(S)$$

holds, where L is a positive constant.

Condition $\mathcal{M}$. Functions $f^i(t, x, y, s)$, $i \in S$, satisfy the following *monotonicity condition* $(\mathcal{M})$ with respect to the functional argument s: for every fixed $t, 0 \leq t \leq T$ and for all functions $s, \tilde{s} \in C_S(\overline{D})$ the following implication:

$$s \overset{(t)}{\leq} \tilde{s} \implies f^i(t, x, y, s) \leq f^i(t, x, y, \tilde{s}) \quad \text{for } x \in \overline{G}, \quad y \in B(S), \quad i \in S$$

holds.

---

[2] Monotonicity of a function $h = h(x)$ is understood in the weak sense, i.e., a function h is increasing or strictly increasing if $x_1 < x_2$ implies $h(x_1) \leq h(x_2)$ or $h(x_1) < h(x_2)$, respectively.

Condition K. We assume that for each $i \in S$, there exists function $k^i = k^i(t, x) > 0$, defined in $\overline{D}$, satisfying the assumption (H_a), and such that the function

$$f_k^i(t, x, y, s) := f^i(t, x, y, s) + k^i(t, x)y^i$$

is increasing with respect to the variable y^i.

Assumption K_κ. We assume that there exists a constant $\kappa \geq 0$ such that for each $i \in S$ the functions

$$f_\kappa^i(t, x, y, s) := f^i(t, x, y, s) + \kappa y^i \quad \text{for arbitrary fixed } i \in S$$

are increasing with respect to the variable y^i.

Then we say that functions $f^i(t, x, y, s)$, $i \in S$, are *quasi-increasing* with respect to y.

Remark 2.3 From condition (L_l^i) it follows that the function

$$f_l^i(t, x, y, s) := f^i(t, x, y, s) + \underline{l}^i(t, x)y^i \quad \text{for arbitrary fixed } i \in S$$

is increasing with respect to the variable y^i. If $\underline{l}^i = \underline{l}^i(t, x)$, $i \in S$, fulfill the assumption (H_a), then this means that the functions f^i, $i \in S$, fulfill the condition (K) with $k^i(t, x) = l^i(t, x)$, $i \in S$.

If $f^i(t, x, y, s)$, $i \in S$, are increasing in y, then the condition (K_κ) is satisfied with $\underline{l}^i \equiv 0$.

If condition (K) holds and $\sup_{S \times \overline{D}} \left|k^i(t, x)\right| = k^0 < \infty$, then condition (K_κ) holds with $\kappa \geq k^0$.

Finally, we remark that from Lipschitz condition (L) with respect to y there follows condition (K_κ); namely the functions $f^i(t, x, y, s), i \in S$, are quasi-increasing with respect to y.

We will say that a domain G has the boundary ∂G of a class $C^{2+\alpha}$ if for every point of ∂G there exists an $(m + 1)$-dimensional neighborhood U such that $U \cap \partial G$ can be represented, for some j $(1 \leq j \leq m)$, in the form

$$x_j = g^j(t, x_1, x_2, \ldots, x_{j-1}, x_{j+1}, \ldots, x_m),$$

where $g^j \in C^{2+\alpha}$.

Moreover, we will assume that:

Assumption H_ϕ. $\phi \in C_S^{2+\alpha}(\Gamma)$, where $0 < \alpha < 1$.

If initial-boundary condition (2.5) is of the form (2.7) and (2.8), i.e.,

$$\phi(t, x) = \begin{cases} \phi_0(x) & \text{for } t = 0, \quad x \in G, \\ h(t, x) & \text{for } (t, x) \in \sigma, \end{cases}$$

then we will assume that:

Assumption $H_{\phi,h}$. $\phi_0 \in C_S^{2+\alpha}(\overline{G})$, $h \in C_S^{2+\alpha}(\sigma)$, where $0 < \alpha < 1$, and compatibility condition

$$h(0, x) = \phi_0(x) \quad \text{for } x \in \partial G$$

holds.

Remark 2.4   If $\phi \in C_S^{2+\alpha}(\Gamma)$ and the boundary $\partial G \in C^{2+\alpha}$, then without loss of generality we can consider the homogeneous initial-boundary condition (2.6), i.e.,

$$z(t, x) = 0 \quad \text{for } (t, x) \in \Gamma$$

for system (2.2) or (2.3).

Indeed, if $\phi(t, x) \not\equiv 0$ on Γ, $\phi \in C_S^{2+\alpha}(\Gamma)$, and $\partial G \in C^{2+\alpha}$, then there exists an extension Φ of the function ϕ onto $\overline{D}$ such that $\Phi \in C_S^{2+\alpha}(\overline{D})$ and

$$\Phi(t, x) = h(t, x) \quad \text{for each } (t, x) \in \Gamma.$$

Let Φ be a given extension of ϕ onto $\overline{D}$ and z be a solution of problem (2.2) and (2.5) in $\overline{D}$, $z \in C_S^{2+\alpha}(\overline{D})$. It is routine to see that the function

$$\overset{*}{z}(t, x) = z(t, x) - \Phi(t, x)$$

satisfies the following homogeneous problem

$$\begin{cases} \mathcal{F}^i[\overset{*}{z}{}^i](t, x) = \overset{*}{f}{}^i\left(t, x, \overset{*}{z}(t, x), \overset{*}{z}\right), & i \in S, \quad \text{for } (t, x) \in D, \\ \overset{*}{z}(t, x) = 0 \quad \text{for } (t, x) \in \Gamma, \end{cases}$$

where

$$\overset{*}{f}{}^i\left(t, x, \overset{*}{z}(t, x), \overset{*}{z}\right) := f^i\left(t, x, \overset{*}{z}(t, x) + \Phi(t, x), \overset{*}{z} + \Phi\right) - \mathcal{F}^i[\Phi^i](t, x).$$

Obviously, the functions $\overset{*}{f}{}^i$ have the same property as the functions f^i.

Accordingly, in what follows, we can confine ourselves to considering homogeneous problem (2.2) and (2.6) in $\overline{D}$ only.

2.8 Lower and Upper Solutions

Functions $u, v \in C_S(\overline{D}) \cap C_S^{1,2}(D) := C_S^{reg}(\overline{D})$ satisfying the infinite systems of inequalities

$$\begin{cases} \mathcal{D}_t u^i(t, x) - \mathcal{A}^i[u^i](t, x) \le f^i(x, u(t, x), u(t, \cdot)) & \text{for } (t, x) \in D, \\ u^i(0, x) \le \phi_0^i(x) \quad \text{for } x \in G, \\ u^i(t, x) \le h^i(x) \quad \text{for } (t, x) \in \sigma \quad \text{and } i \in S \end{cases} \tag{2.20}$$

and

$$\begin{cases} \mathcal{D}_t v^i(t, x) - \mathcal{A}^i[v^i](t, x) \geq f^i(x, v(t, x), v(t, \cdot)) & \text{for } (t, x) \in D, \\ v^i(0, x) \geq \phi_0^i(x) & \text{for } x \in G, \\ v^i(t, x) \geq h^i(x) & \text{for } (t, x) \in \sigma, \text{ and } i \in S \end{cases} \tag{2.21}$$

are called, respectively, a *lower solution* and an *upper solution* of problem (2.2) and (2.7)–(2.9) in $\overline{D}$.

We define an ordered pair of a lower solution u and an upper solution v of problem (2.2) and (2.7)–(2.9) in $\overline{D}$ when $u(t, x) \leq v(t, x)$ in $\overline{D}$.

We notice that the inequality $u \leq v$ in $\overline{D}$ does not follow directly from the inequalities (2.20) and (2.21). Therefore we adopt the following assumption A_0.

In addition to the problem (2.30)–(2.32) in $\overline{D}$ we consider the corresponding infinite systems of equations of the form

$$-\mathcal{A}^i[Z^i](x) = f^i(x, Z(x), Z), \quad i \in S, \quad \text{for } x \in G, \tag{2.22}$$

where $Z := Z(\cdot)$ and $Z = \{Z^i\}_{i \in S}$ denote an element of the space $L_S^p(G)$ of functions

$$Z: S \times \overline{G} \to \mathbb{R}, \quad (i, x) \mapsto Z^i(x).$$

For (2.22) we consider the boundary value problem with the *Dirichlet* condition

$$Z(x) = h(x) \quad \text{for } x \in \partial G. \tag{2.23}$$

Assumption A_0. We assume that there exists at least one ordered pair u_0 and v_0 of a lower and an upper solution of problem (2.2), (2.7)–(2.9) in $\overline{D}$, and $u_0, v_0 \in C_S^{0+\alpha}(\overline{D})$.

A function Z is called a *regular function* in $\overline{G}$ if

$$Z \in C_S^{\text{reg}}(\overline{G}) := C_S(\overline{G}) \cap C_S^2(G).$$

Functions $U, V \in C_S^{\text{reg}}(\overline{G})$ satisfying the infinite systems of inequalities

$$\begin{cases} -\mathcal{A}^i[U^i](x) \leq f^i(x, U(x), U) & \text{for } x \in G \text{ and } i \in S, \\ U(x) \leq h(x) & \text{for } x \in \partial G \end{cases} \tag{2.24}$$

and

$$\begin{cases} -\mathcal{A}^i[V^i](x) \geq f^i(x, V(x), V) & \text{for } x \in G \text{ and } i \in S, \\ V(x) \geq h(x) & \text{for } x \in \partial G \end{cases} \tag{2.25}$$

are called, respectively, *a lower solution* and *an upper solution* of problem (2.22) and (2.23) in $\overline{G}$.

A pair U and V of a lower and an upper solution of problem (2.22) and (2.23) in $\overline{G}$ is called an ordered pair if $U(x) \leq V(x)$ in $\overline{G}$.

Remark 2.5 Inequality $U \le V$ does not follow directly from the definitions of a lower and an upper solution, i.e., from inequalities (2.24) and (2.25). There are examples of a lower function greater than an upper solution. In this case the problem of the existence of solutions cannot be solved by an iterative method. In the case of $V_0 \le U_0$ the iterative method breaks down, because starting an iteration process from V_0 (resp. U_0) we get a decreasing (resp. increasing) sequence.

Example Consider the following equation:

$$u'' + \lambda u e^{-u} = 0 \quad \text{for } -1 \le x \le 1$$

with the boundary conditions $u(-1) = u(1) = 0$. Obviously, $u_1(x) = 0$ is a solution of this problem and it is an upper solution. But $u_2(x) = C_1 \cos(\pi x/2) \ge 0$ for $\lambda > \pi^2/4$ and $C_1 = ln(4\lambda/\pi^2)$ is a lower solution because $u_2'' + \lambda u_2 e^{-u_2} \ge 0$.

Therefore, we will adopt the following assumption:

Assumption A_0^*. We assume that there exists at least one ordered pair U_0 and V_0 of lower and upper solutions of problem (2.22) and (2.23) in $\overline{G}$, i.e.,

$$U_0(x) \le V_0(x) \quad \text{for } x \in \overline{G}, \tag{2.26}$$

and $U_0, V_0 \in C_S^{0+\alpha}(\overline{G})$.

Remark 2.6 We remark that every ordered pair of lower and upper solutions U_0 and V_0 of problem (2.22) and (2.23) is an ordered pair of lower and upper solutions of problem (2.2) and (2.7)–(2.9) when $\phi_0 \in \langle U_0, V_0 \rangle$.

For an ordered pair U_0 and V_0 we define the *sector* $\langle U_0, V_0 \rangle$ in the space $L_S^p(G)$ as follows:

$$\langle U_0, V_0 \rangle := \left\{ s \in L_S^p(G) \colon U_0(x) \le s(x) \le V_0(x) \text{ for } x \in \overline{G} \right\}, \tag{2.27}$$

and the *sets*

$$\mathcal{K}_e := \left\{ (x, y, s) \colon x \in \overline{G}, \ y \in \langle \underline{m}, \overline{M} \rangle, \ s \in \langle U_0, V_0 \rangle \right\} \tag{2.28}$$

and

$$K^* := \left\{ (t, x, s) \colon (t, x) \in \overline{G}, s \in \langle U_0, V_0 \rangle \right\}, \tag{2.29}$$

where

$$\underline{m}^i = \inf_{\overline{D}} U_0^i(x), \quad \underline{m} = \left\{ \underline{m}^i \right\}_{i \in S},$$
$$\overline{M}^i = \sup_{\overline{D}} V_0^i(x), \quad \overline{M} = \left\{ \overline{M}^i \right\}_{i \in S}.$$

We recall that from Assumption A_0^* there follows that the assumptions on the functions f^i may be weakened to hold locally in the set $\mathcal{K}_e$. Therefore, all of our theorems will be true locally only within the sector $\langle U_0, V_0 \rangle$.

2.9 Stability of Solutions of Infinite Systems

We consider an infinite *system* of weakly coupled *autonomous*[3] equations of the form

$$\mathcal{D}_t z^i(t, x) - \mathcal{A}^i[z^i](t, x) = f^i(x, z(t, x), z(t, \cdot)), \quad i \in S, \tag{2.30}$$

where

$$\mathcal{A}^i := \sum_{j,k=1}^m a^i_{jk}(x)\mathcal{D}^2_{x_j x_k}$$

for $(t, x) \in D := (0, T] \times G$ where T is an arbitrary positive constant, $x = (x_1, \ldots, x_m) \in G \subset \mathbb{R}^m$, G is an open and bounded domain with a $C^{2+\alpha}$ $(0 < \alpha < 1)$ boundary, the operators $\mathcal{A}^i$, $i \in S$, are uniformly elliptic in $\overline{G}$ and S is an arbitrary set of indices (finite or infinite).

For a fixed $t \geq 0$, by $z(t, \cdot)$ we denote an element of the space $C_S(\overline{G})$ of continuous functions from the closure of domain G into $\mathcal{B}(S)$

$$z(t, \cdot): \overline{G} \to \mathcal{B}(S), \quad x \mapsto z(t, x).$$

We shall assume[4] that $p > m$, and

$$f^i: \overline{G} \times \mathcal{B}(S) \times L^p_S(G) \to \mathbb{R}, \quad (x, y, s) \mapsto f^i(x, y, s), \quad i \in S$$

are given functions.

We assume equations to be autonomous to be able to study the asymptotic behavior of solutions of the systems considered.

For system (2.30), we consider the problem with the initial condition

$$z(0, x) = \phi_0(x) \quad \text{for } x \in G \tag{2.31}$$

and the boundary condition

$$z(t, x) = h(x) \quad \text{for } (t, x) \in \sigma, \tag{2.32}$$

where the compatibility condition

$$h(x) = \phi_0(x) \quad \text{for } x \in \partial G \tag{2.33}$$

holds.

---

[3] A *system* is called *autonomous* if its right-hand sides do not depend evidently on time.
[4] Under this assumption we can apply the Rellich-Kondrachov theorem on compact imbeddings.

If $Z = Z(x)$ is a solution of problem (2.22) and (2.23), then we shall call Z a *stable solution of problem* (2.30)–(2.33) if, for any $\varepsilon > 0$, there exists a $\delta = \delta(\varepsilon) > 0$ such that

$$\|Z(\cdot) - \phi_0(\cdot)\|_{C_S(\overline{G})} < \delta \quad \text{implies} \quad \|Z(\cdot) - z(t, \cdot)\|_{C_S(\overline{G})} < \varepsilon$$

for each $t > 0$, where $z = z(t, x)$ is a solution of problem (2.30)–(2.33).

We shall call a solution $Z = Z(x)$ *asymptotically stable solution of problem* (2.30)–(2.33) if it is a stable solution of this problem and

$$\lim_{t \to \infty} \|Z(\cdot) - z(t, \cdot)\|_{C_S(\overline{G})} = 0.$$

CHAPTER 3

Differential Inequalities

3.1 Introduction

Comparison theorems play a very important role as differential inequalities methods for finite and infinite systems of equations. These methods are commonly used in numerical methods for the computation of solutions and to prove the continuous dependence of the solutions upon the right-hand side of the system and upon the initial and boundary conditions. Comparison theorems can be used to prove the existence and uniqueness of solutions of finite and infinite systems of equations. We begin this chapter with the theorems on the continuous dependence of solutions of the Fourier or/and Cauchy problems upon initial and boundary conditions, and the right-hand sides of the systems.

3.2 Comparison Theorems for Finite Systems

Let us consider a weakly coupled finite system of equations of the form

$$\mathcal{F}^q[z^q](t, x) = f^q(t, x, z(t, x), z), \quad q = 1, 2, \ldots, r, \tag{3.1}$$

where

$$\mathcal{F}^q := \mathcal{D}_t - \mathcal{L}^q, \quad \mathcal{L}^q := \sum_{j,k=1}^{m} a_{jk}^q(t, x)\mathcal{D}_{x_j x_k}^2 - \sum_{j=1}^{m} b_j^q(t, x)\mathcal{D}_{x_j}, \tag{3.2}$$

$(t, x) = (t, x_1, \ldots, x_m) \in (0, T) \times G := D, T \leq \infty, G \subset \mathbb{R}^m, G$ is an open and bounded domain with boundary $\partial G \in C^{2+\alpha}$ ($0 < \alpha < 1$).

Directly from Szarski's theorem on weak differential inequalities for finite systems, there follows:

Theorem 3.1 *Let functions $f^q(t, x, y, s), q = 1, 2, \ldots, r$, be defined for*
$(t, x, y, s) \in \overline{D} \times \mathbb{R}^r \times C_r(\overline{D})$ *and operators $\mathcal{F}^q, q = 1, 2, \ldots, r$, be uniformly parabolic in*
$\overline{D}$. *Assume that:*

Mathematical Neuroscience. http://dx.doi.org/10.1016/B978-0-12-411468-5.00003-X

i. *the functions $f^q(t, x, y, s), q = 1, 2, \ldots, r$, satisfy conditions* **(W+)** *and* **(L₁)** *with respect to y and satisfy conditions* **(W)**, **(L)**, **(V)** *with respect to s;*

ii. *the functions $u, v \in C_r^{\text{reg}}(\overline{D})$ fulfill the following finite systems of inequalities:*

$$\mathcal{F}^q[u^q](t, x) \leq f^q(t, x, u(t, x), u) \quad \text{for } (t, x) \in D \quad \text{and} \quad q = 1, 2, \ldots, r$$

and

$$\mathcal{F}^q[v^q](t, x) \geq f^q(t, x, v(t, x), v) \quad \text{for } (t, x) \in D \quad \text{and} \quad q = 1, 2, \ldots, r.$$

iii. *Finally, suppose that*

$$u(0, x) \leq v(0, x) \quad \text{for } x \in G$$

and

$$u(t, x) \leq v(t, x) \quad \text{for } (t, x) \in \sigma.$$

Under these assumptions, there is

$$u(t, x) \leq v(t, x) \quad \text{for } (t, x) \in \overline{D}.$$

3.3 Maximum Principles for Finite Systems

As a corollary to this theorem we obtain the strong maximum principle.

Theorem 3.2 (Maximum principle)   *Let us consider the finite system of parabolic differential-functional inequalities of the form*

$$\mathcal{F}^q[u^q](t, x) \leq f^q(t, x, u(t, x), u) \quad \text{for } (t, x) \in D, \quad q = 1, 2, \ldots, r, \qquad (3.3)$$

where the functions $f^q(t, x, y, s), q = 1, 2, \ldots, r$, are defined for $(t, x, y, s) \in \overline{D} \times \mathbb{R}^r \times C_r(\overline{D})$ and the diagonal operators $\mathcal{F}^q, q = 1, 2, \ldots, r$, are uniformly parabolic in $\overline{D}$.

Assume that:

i. *the functions $f^q(t, x, y, s), q = 1, 2, \ldots, r$, satisfy conditions* **(W+)** *and the right-hand side Lipschitz condition with respect to y and conditions* **(W)**, **(L)** *and* **(V)** *with respect to s;*

ii. *suppose u is a regular solution of (3.3) in $\overline{D}$, satisfying inequalities*

$$u(0, x) \leq c \quad \text{for } x \in G,$$
$$u(t, x) \leq c \quad \text{for } (t, x) \in \sigma,$$

where $c = \{c^q\}_{q=1,2,\ldots,r} = const.$

iii. *Finally, assume that*

$$f^q(t, x, c, c) \leq 0 \quad for\ (t, x) \in D \quad and \quad q = 1, 2, \ldots, r.$$

Under these assumptions, there is

$$u(t, x) \leq c \quad for\ (t, x) \in \overline{D}.$$

As a direct consequence of this theorem we get the following lemmas:

Lemma 3.1 (The Positivity Lemma) *Let $u \in CB(\overline{D})$ (a class of continuous and bounded functions) and suppose that*

$$\begin{aligned}
\mathcal{F}^q[u^q](t, x) &\geq 0 \quad for\ (t, x) \in D, \\
u^q(0, x) &\geq 0 \quad for\ x \in G, \\
u^q(t, x) &\geq 0 \quad for\ (t, x) \in \sigma \quad and \quad q = 1, 2, \ldots, r.
\end{aligned}$$

Then $u(t, x) \geq 0$ for $(t, x) \in \overline{D}$.

Lemma 3.2 (The Gronwall-Bellman Inequality) *Let u and f be continuous and nonnegative functions defined on $[\alpha, \beta] \subset \mathbb{R}$, and let c be a nonnegative constant. Then the inequality*

$$u(t) \leq c + \int_\alpha^t f(s)u(s)ds \quad for\ t \in [\alpha, \beta],$$

implies that

$$u(t) \leq c + \exp\left(\int_\alpha^t f(s)u(s)ds\right) \quad for\ t \in [\alpha, \beta].$$

Lemma 3.3 *Let a set D have property $\mathcal{P}$ and $\bar{D} \subset D$. Let the real function*

$$f : D \times \mathbb{R} \times \mathbb{R}^m \times \mathcal{Q}_{m \times m} \to \mathbb{R},\ (t, x, y, p, q) \mapsto f(t, x, y, p, q)$$

i. *satisfy the Lipschitz condition with respect to y, p and q with a positive constant L in the following form*

$$[f(t, x, y, p, q) - f(t, x, \tilde{y}, \tilde{p}, \tilde{q})]sgn(y - \tilde{y})$$
$$\leq L\left(|y - \tilde{y}| + \sum_{j=1}^m |p_j - \tilde{p}_j| + \sum_{j,k=1}^m |q_{jk} = \tilde{q}_j k|\right).$$

ii. *Let $u : D_0 \to \mathbb{R}$ be a regular, bounded and parabolic solution of equation*

$$\mathcal{D}_t u(t, x) = f(t, x, u(t, x), \mathcal{D}_x u(t, x), \mathcal{D}^2_{xx} u(t, x)) \quad for\ (t, x) \in D.$$

iii. *Suppose* $v : D_0 \to \mathbb{R}$ *is a regular and bounded solution of inequality*

$$|\mathcal{D}_t v(t, x) - f((t, x, v(t, x), \mathcal{D}_x v(t, x), \mathcal{D}^2_{xx} v(t, x))| \leq \quad for \ (t, x) \in D$$

  with a positive constant M,
iv. *satisfying the initial and boundary inequalities*

$$|v(t, x) - u(t, x)| \leq \varepsilon \quad for \ (t, x) \in D_0 \backslash D.$$

  Under these assumptions, we have the error estimate

$$|v(t, x) - u(t, x)| \leq (\varepsilon + ML^{-1}) \exp Lt - ML^{-1} \quad for \ (t, x) \in D.$$

These lemmas play a fundamental role in applying maximum principles to problems arising in neuroscience.

3.4 Comparison Theorems for Infinite Systems

Based on lemma 3.3 the following comparison theorem for infinite systems of equations is proved.

Theorem 3.3 *Let the functions*

$$f^i : D \times \mathcal{B}(S) \times \mathbb{R}^m \times \mathbb{Q}_{m \times m} \times \mathcal{Z} \to \mathbb{R}, \quad (t, x, y, p, q, s) \mapsto f^i(t, x, y, p, q, s), i \in S,$$

i. *satisfy the Lipschitz condition with respect to y, p, q, and the Lipschitz-Volterra condition with respect to the functional argument s in the following form*

$$[f^i(t, x, y, p, q, s) - f^i(t, x, \tilde{y}, \tilde{p}, \tilde{q}, \tilde{s})]sgn(y^i - \tilde{y}^i)$$
$$\leq L(\|y - \tilde{y}\|_{\mathcal{B}(S)} + \sum_{j=1}^m |p_j - \tilde{p}_j| + \sum_{j,k=1}^m |q_{jk} - \tilde{q}_jk| + \|s - \tilde{s}\|_t)$$

  with a positive constant L.
ii. *Let $u \in \mathcal{Z}$ be a regular and parabolic solution of the system*

$$\mathcal{D}_t u^i(t, x) = f^i(t, x, u(t, x), \mathcal{D}_x u^i(t, x), \mathcal{D}^2_{xx} u^i(t, x), u), \quad i \in S,$$

  for $(t, x) \in D$ and
iii. $v \in \mathcal{Z}$ *be a regular solution of the system of inequalities for a given $\varepsilon \geq 0$*

$$|\mathcal{D}_t v^i(t, x) - f^i(t, x, v(t, x), \mathcal{D}_x v^i(t, x), \mathcal{D}^2_{xx} v^i(t, x), v)| \leq \varepsilon, \quad i \in S,$$

  for $(t, x) \in D$;

iv. *u and v satisfy the initial and boundary inequalities*

$$|v^i(t, x) - u^i(t, x)| \le \varepsilon, \quad i \in S, \quad for (t, x) \in D_0 \backslash D.$$

Under these assumptions the following inequality holds true

$$||u - v||_0 \le \varepsilon A^P \quad for (t, x) \in D, \tag{3.4}$$

where ([1])

$$h := L^{-1} \ln (5/4), \quad P := \left[\frac{T}{h}\right] + 1, A := (5 + L^{-1})2^{-1}.$$

Let $T > 0$ and layer given by $\Omega = \{(t, x) : t \in (0, T), x \in \mathbb{R}^m\}$ and S be an infinite set of indices. Let $B(S)$ be the space of bounded mappings $v : S \ni i \mapsto v^i \in \mathbb{R}$ with the supremum norm

$$||v||_{B(S)} := \sup\{|v^i| : i \in S\}.$$

For every nonempty set $X \subset \mathbb{R}^m$ we denote by $C_S(X)$ the space of mappings

$$w : X \ni x \mapsto w(x) \in B(S), \quad where \ w(x) : S \ni i \mapsto w^i(x) \in \mathbb{R},$$

and the functions w^i are continuous in X. For w use the notation $w = \{w^i\}_{i \in S}$ as well.

Theorem 3.4 *Let $f := \{f^i\}_{i \in S}$ and $\varphi := \{\varphi^i\}_{i \in S}$ be given,*

$$f^i : \overline{\Omega} \times B(S) \times C_S(\Omega) \to \mathbb{R},$$
$$\varphi^i : \mathbb{R}^m \to \mathbb{R}, \quad i \in S.$$

Let $u = \{u^i\}_{l \in S}$ where each u^i is an unknown function of the variables $(t, x) = (t, x_1, \ldots, x_m)$, and set

$$\mathcal{F}^i := \mathcal{D}_t - L^i, \quad L^i := \sum_{j,k=1}^m a^i_{jk}(t, x)\mathcal{D}^2_{x_j x_k} + \sum_{j=1}^m b^i_j(t, x)D_{x_j},$$

where $A^i_{jk}(t, x) = A^i_{j=kj}(t, x)$ for $(t, x) \in \Omega$, $j, k = 1, 2, \ldots, m$, $i \in S$.

We consider an infinite system of the form

$$\mathcal{F}^i[u^i](t, x) = f^i(t, x, u(t, x), u), \quad i \in S, \tag{3.5}$$

together with the initial condition

$$u(0, x) = \varphi_0(x) \quad for \ x \in \mathbb{R}^m. \tag{3.6}$$

---

[1] For a positive real number a, we denote by $[a]$ the greatest integral not exceeding a.

Let $f^i = f^i(t, x, s, p)$ be Lipschitz continuous in s and in p (uniformly with respect to $i \in S$), i.e.,

$$|f^i(t, x, s, p) - f^i(t, x, \tilde{s}, \tilde{p})| \leq L_1 \|s - \tilde{s}\|_{B(S)} + L_2 \|p - \tilde{p}\|_{0,t}.$$

If $v, w \in C_S(\overline{\Omega})$ satisfy

$$\mathcal{F}^i[v^i](t, x) = f^i(t, x, v(t, x), v),$$
$$\mathcal{F}^i[w^i](t, x) = \overline{f}^i(t, x, w(t, x), w), \quad i \in S, \tag{3.7}$$

and there exists a nonnegative constant M independent of $i \in S$ such that of the Lipschitz continuity of f^i in s and p we have

$$\left| f^i(t, x, v(t, x), v) - \overline{f}^i(t, x, w(t, x), w) \right| \leq M \tag{3.8}$$

then

$$\|v - w\|_{0,t} \leq C \|v(0, \cdot) - w(0, \cdot)\|_0 \, e^{tCL} + \int_0^t CM e^{(t-\tau)CL} d\tau$$

provided $v - w \in CB_S(\overline{\Omega})$, where $L = L_1 + L_2$.

Proof.   Set $\tilde{z}(t) := \|z(t)\|_{B(S)}$, where $z(t) = \{z^i(t)\}_{i \in S}$ and

$$z^i(t) = \sup_{x \in \mathbb{R}^m, \bar{t} \leq t} |v^i(\bar{t}, x) - w^i(\bar{t}, x)| = \|v^i - w^i\|_{0,t}.$$

Since v, w satisfy (3.7), making use of (3.8)

$$v^i(t, x) - w^i(t, x)$$
$$= \int_{\mathbb{R}^m} \Gamma^i(t, x; 0, \xi)[v^i(0, \xi) - w^i(0, \xi)]d\xi$$
$$\quad + \int_0^t \int_{\mathbb{R}^m} \Gamma^i(t, x; 0, \xi)[f^i(\tau, \xi, v(\tau, \xi), v) - \overline{f}^i(\tau, \xi, w(\tau, \xi), w)]d\xi d\tau$$
$$\leq Cz^i(0) + \int_0^t \int_{\mathbb{R}^m} \Gamma^i(t, x; 0, \xi)|f^i(\tau, \xi, v(\tau, \xi), v) - f^i(\tau, \xi, v(\tau, \xi), v)|d\xi d\tau$$
$$\quad + \int_0^t \int_{\mathbb{R}^m} \Gamma^i(t, x; 0, \xi)|f^i(\tau, \xi, v(\tau, \xi), v) - f^i(\tau, \xi, w(\tau, \xi), w)|d\xi d\tau$$
$$\quad + \int_0^t \int_{\mathbb{R}^m} \Gamma^i(t, x; 0, \xi)|f^i(\tau, \xi, w(\tau, \xi), w) - \overline{f}^i(\tau, \xi, w(\tau, \xi), w)|d\xi d\tau$$
$$\leq Cz^i(0) + \int_0^t C(L_1 \|v(t, x) - w(t, x)\|_{B(S)} + L_2 \|v - w\|_{0,\tau} + M)d\tau$$
$$\leq Cz^i(0) + \int_0^t C[(L_1 + L_2)\|z(\tau)\|_{B(S)} + M]d\tau, \quad i \in S.$$

Therefore $\tilde{z}$ satisfies the integral inequality

$$\tilde{z}(t) \leq C\tilde{z}(t) + \int_0^t C[M + (L_1 + L_2)\tilde{z}(\tau)]d\tau.$$

Thus, the Gronwall-Bellman lemma yields

$$\tilde{z}(t) \leq C\tilde{z}(t)e^{tC(L_1+L_2)} + \int_0^t CMe^{(t-\tau)C(L_1+L_2)}d\tau.$$

This completes the proof. □

Theorem 3.5 *Let assumption $\mathcal{H}_t$ hold. Let $f^i = f^i(t, x, s, p)$ be increasing in s and p and satisfy, uniformly with respect to $i \in S$, the one-sided Lipschitz condition with respect to s and p, i.e.,*

$$f^i(t, x, s, p) - f^i(t, x, \tilde{s}, \tilde{p}) \leq L_1 \|s - \tilde{s}\|_{B(S)} + L_2 \|p - \tilde{p}\|_{0,t} \quad for \; s \geq \tilde{s}, \quad p \geq \tilde{p}.$$

If $v, w \in CB_S(\overline{\Omega})$ satisfy

$$\begin{aligned}
\mathcal{F}^i[v^i](t, x) &= f^i(t, x, v(t, x), v), \\
\mathcal{F}^i[w^i](t, x) &= \overline{f}^i(t, x, w(t, x), w), \quad i \in S,
\end{aligned} \tag{3.9}$$

then the initial inequality $v(0, x) \leq w(0, x)$ carries over to $\overline{\Omega}$.

Proof. Set $y(t) = \{y^i(t)\}_{i \in S}$ where $y^i(t) := \max\{0, z^i(t)\}$ and

$$z^i(t) = \sup_{x \in \mathbb{R}^m, \tilde{t} \leq t} [v^i(\tilde{t}, x) - w^i(\tilde{t}, x)].$$

It is obvious that $0 \leq y^i(t) < \infty$ and $z^i(t) \leq y^i(t)$ for all $i \in S$. Let $\mathbb{I} \in CB_S(\overline{\Omega})$ be the function whose every component is the constant function equal to 1.

Now, the Lipschitz condition and the fact that f^i, $i \in S$, are increasing in s and p yield

$$\begin{aligned}
f^i(t, x, v(t, x), v) &- f^i(t, x, w(t, x), w) \\
&= f^i(t, x, [w + (v - w)](t, x), [w + (v - w)]) - f^i(t, x, w(t, x), w) \\
&\leq f^i(t, x, w(t, x) + y\mathbb{I}(t, x), w + y\mathbb{I}) - f^i(t, x, w(t, x), w) \\
&\leq L_1 \|y(t)\|_{B(S)} + L_2 \|y\mathbb{I}\|_{0,t} = (L_1 + L_2) \|y(t)\|_{B(S)}.
\end{aligned}$$

Since v, w satisfy (3.9), for all $i \in S$ we have

$$\begin{aligned}
v^i(t, x) - w^i(t, x) &= \int_{\mathbb{R}^m} \Gamma^i(t, x; 0, \xi)[v^i(0, \xi) - w^i(0, \xi)]d\xi \\
&\quad + \int_0^t \int_{\mathbb{R}^m} \Gamma^i(t, x; \tau, \xi)[f^i(\tau, \xi, v(\tau, \xi), v) - f^i(\tau, \xi, w(\tau, \xi), w)]d\xi d\tau \\
&\leq \int_0^t C(L_1 + L_2) \|y(\tau)\|_{B(S)} d\tau.
\end{aligned}$$

Thus, the function $y(t)$ is equal to 0. This completes the proof. □

3.5 Infinite Systems of Nonlinear Differential Inequalities

Let us consider the weakly coupled infinite system of nonlinear differential inequalities of the form

$$\mathcal{D}_t z^i(t, x) < f^i\left(t, x, z(t, x), \mathcal{D}_x z^i(t, x), \mathcal{D}^2_{xx} z^i(t, x), z\right), \quad i \in S, \tag{3.10}$$

in a domain $D \subset \mathbb{R}^{m+1}$, where the ith equation contains derivatives of one unknown function z^i only, S is an arbitrary set of indices (finite or infinite), z stands for the mapping

$$z: S \times \overline{D} \to \mathbb{R}, \quad (i, t, x) \mapsto z^i(t, x)$$

composed of unknown functions z^i, and $f^i, i \in S$, are given functions

$$f^i: D \times \mathcal{B}(S) \times \mathbb{R}^m \times \mathcal{Q}_{m \times m} \times \mathcal{Z} \to \mathbb{R}, \quad (t, x, y, p, q, s) \mapsto f^i(t, x, y, p, q, s), \quad i \in S,$$

where $\mathcal{Z}$ denotes some subset (not necessarily a linear subspace) of the space $C_S(\overline{D})$. Thus the ith equation contains derivatives of one unknown function z^i only and the right-hand sides f^i of this system are functionals of z.

We say that a *function* $w \in C_S(\overline{D})$ belongs to the *class* $C_S^*(D)$ if it possesses continuous first spatial derivatives $\mathcal{D}_x w^i(t, x)$ for $i \in S$, in D.

3.6 Ellipticity and Parabolicity of Nonlinear Inequalities

In this section we present the definition of ellipticity and parabolicity of solutions of infinite systems of nonlinear differential inequalities.

Definition of ellipticity in Szarski's sense. Suppose that the function $u = u(t, x) \in C_S^*(D)$ is given. We say that the function $f^i(t, x, y, p, q, s)$, for arbitrary fixed $i \in S$, is elliptic with respect to the function $u = u(t, x)$ at the point $(t, x) \in D$ if for any two real symmetric matrices $q, \tilde{q} \in \mathcal{Q}_{m \times m}, q = (q_{jk})_{j,k=1,...,m}$ and $\tilde{q} = (\tilde{q}_{jk})_{j,k=1,...,m}$ such that $\tilde{q} \geq q^2$ there is

$$f^i(t, x, u(t, x), \mathcal{D}_x u^i(t, x), \tilde{q}, u) \geq f^i(t, x, u(t, x), \mathcal{D}_x u^i(t, x), q, u).$$

If the above property holds true for every point $(t, x) \in \mathcal{D}$, then we say that $f^i(t, x, y, p, q, s)$ is elliptic with respect to u in D.

---

[2] The inequality $\tilde{q} \geq q$ means that the quadratic form in $\xi_1, \xi_2, \ldots, \xi_m$ is positive, i.e.,

$$\sum_{j,k=1}^{m} (\tilde{q}_{jk} - q_{jk}) \xi_j \xi_k \geq 0 \quad \text{for all } \xi = (\xi_1, \xi_2, \ldots, \xi_m) \in \mathbb{R}^m.$$

Suppose for example that we have the linear equation

$$\mathcal{D}_t u(t, x) = \sum_{j,k=1}^{m} a_{jk}(t, x) \mathcal{D}_{x_j x_k} u(t, x) + \sum_{j=1}^{m} b_j(t, x) \mathcal{D}_{x_j} u(t, x)$$
$$+ c(t, x) u(t, x) + f(t, x) \quad \text{for } (t, x) \in D,$$

which is parabolic at a point $(t, x) \in D$. This means that the quadratic form in $\xi_1, \xi_2, \ldots, \xi_m$ is positive in $(t, x) \in D$, i.e.,

$$\sum_{j,k=1}^{m} a_{jk}(t, x) \xi_j \xi_k \geq 0 \quad \text{for } \xi = (\xi_1, \xi_2, \ldots, \xi_m) \in \mathbb{R}^m \quad \text{(here and below)}.$$

Therefore the right-hand member

$$f(t, x, u, p, q) = \sum_{j,k=1}^{m} a_{jk}(t, x) q_{jk} + \sum_{j=1}^{m} b_j(t, x) p_j + c(t, x) u + f(t, x)$$

of equation at a point (t, x) is elliptic at (t, x) with respect to any function $u = u(t, x) \in C_S^*(D)$. Here the equation is parabolic at (t, x).

Definition of parabolicity in Szarski's sense. Let us consider the infinite system of (3.10) and for this system we consider the initial-boundary condition

$$z(t, x) - \phi(t, x) \quad \text{for } (t, x) \in D_0 \backslash D, \tag{3.11}$$

where $\phi = \{\phi^i\}_{i \in S}$ is a given function $\phi \colon D_0 \backslash D \to \mathcal{B}(S)$.

A mapping $w \in C_S(D_0)$ is called a *regular solution* of problem (3.10) and (3.11) if the derivatives $\mathcal{D}_t w^i(t, x)$, $\mathcal{D}_x w^i(t, x)$, $\mathcal{D}_{xx}^2 w^i(t, x)$ are continuous in D, system (3.10) is satisfied at each point D, and initial-boundary condition (3.11) holds on $D_0 \backslash D$.

The regular solution z of system (3.10) is called a *regular parabolic solution of* (3.10) *in D* if all the functions $f^i, i \in S$, are elliptic functions with respect to this solution z in D.

Let us consider the nonlinear infinite system of equations of the form (3.9) from Theorem 3.3, one can obtain the following uniqueness criterion for the solution of the first initial-boundary value problem.

Theorem 3.6 (Uniqueness criterion) *Let us suppose that assumption (i) of Theorem 3.3 holds true. Then the system of equations of the form (3.9) admits at most one regular and parabolic solution $u \in \mathcal{Z}$ satisfying the initial-boundary condition (3.10).*

Proof. Suppose there exists a second regular solution (not necessarily parabolic) of our problem. Let us denote it by v. All assumptions of Theorem 3.3 are now satisfied with $\varepsilon = 0$, and therefore by (3.24) we get $u = v$ in D_0. This completes the proof. $\qquad \square$

Definition of ellipticity in Besala's sense. Given a function $u = u(t, x)$ of class $C_S^*(D)$ in the domain $\overline{D}$, the *function* $f^i(t, x, y, p, q, s)$ for arbitrary fixed i, $i \in S$, is said to be *uniformly elliptic with respect to u* in D in Besala's sense if there exists a constant $\kappa > 0$ such that for all $(t, x) \in D$ and any two real symmetric matrices $q, \tilde{q} \in \mathcal{Q}_{m \times m}, q = (q_{jk})$ and $\tilde{q} = (\tilde{q}_{jk}), j, k = 1, 2, \ldots, m$, there is

$$\tilde{q} \geq q \Rightarrow f^i\left(t, x, u(t, x), D_x u^i(t, x), \tilde{q}, u\right) - f^i\left(t, x, u(t, x), D_x u^i(t, x), q, u\right)$$

$$\geq \kappa \sum_{j=1}^m (\tilde{q}_{jj} - q_{jj}). \tag{3.12}$$

A *system* of differential inequalities (3.10) is called *uniformly parabolic with respect to the function* $u \in C_S^*(D)$ in $\overline{D}$ if all the functions $f^i, i \in S$, are uniformly elliptic with respect to this function u in $\overline{D}$.

In particular, if (3.12) is satisfied with $\kappa = 0$, then f^i is called *parabolic in Szarski's sense with respect to the function* u in D. The uniform parabolicity in Besala's sense implies the parabolicity in Szarski's sense.

Remark 3.1   Let us consider the infinite system of semilinear differential inequalities of form (3.10). If the operators $\mathcal{F}^i, i \in S$, are *uniformly parabolic* in $\overline{D}$, then these operators are uniformly parabolic in $\overline{D}$ with respect to any function $u \in C_S^*(\overline{D})$.

On the other hand, it can easily be shown that if f^i is of class C^1 with respect to variables $q = (q_{jk})_{j,k=1,2,\ldots,m}$, then the condition

$$\sum_{j,k=1}^m D_{q_{jk}} f^i(t, x, y, p, q, s) \xi_j \xi_k \geq \lambda |\xi|^2 \quad \text{for all } \xi \in \mathbb{R}^m$$

with a constant $\lambda > 0$, implies the uniform parabolicity in Besala's sense.

3.7 Weak Differential Inequalities for Infinite Systems

We formulate a theorem on weak differential inequalities which generalizes Theorem 3.1.

Theorem 3.7   *Suppose that the subset $\mathcal{Z}$ of the space $C_S(\overline{D})$ has the following property: for any $w \in \mathcal{Z}$ and for any nonnegative constant c the mapping*

$$w - c\colon S \times \overline{D} \to \mathbb{R}, \quad (i, t, x) \mapsto w^i(t, x) - c$$

belongs to $\mathcal{Z}$.

Let the functions

$$f^i\colon D \times \mathbb{R} \times \mathbb{R}^m \times \mathcal{Q}_{m \times m} \times \mathcal{Z} \to \mathbb{R}, \quad (t, x, y, p, q, s) \mapsto f^i(t, x, y, p, q, s), \quad \text{for } i \in S,$$

i. *satisfy the following monotonicity condition form with respect to s:*

$$(\mathcal{M}) \quad s \overset{(t)}{\leq} \tilde{s} \Rightarrow f^i(t, x, y, p, q, s) \leq f^i(t, x, y, p, q, \tilde{s}), \quad i \in S;$$

ii. *and the Lipschitz-Volterra condition of the following form*

$$[f^i(t, x, y, p, q, s) - f^i(t, x, \tilde{y}, \tilde{p}, \tilde{q}, \tilde{s})] \mathrm{sgn}(y - \tilde{y})$$

$$\leq L \left(|y - \tilde{y}| + \sum_{j=1}^{m} |p_j - \tilde{p}_j| + \sum_{j,k=1}^{m} |q_{jk} - \tilde{q}_{jk}| + [s - \tilde{s}]_t \right), \quad i \in S$$

for $s \geq \tilde{s}$, with some positive constant L.

iii. *Assume the mapping $u \in Z$ to be a regular and parabolic solution of the system*

$$\mathcal{D}_t u^i(t, x) \leq f^i\left(t, x, u^i(t, x), \mathcal{D}_x u^i(t, x), \mathcal{D}_{xx} u^i(t, x), u\right) \quad for \ (t, x) \in D, \quad i \in S.$$

iv. *Let the mapping $v \in Z$ be a regular solution of the system*

$$\mathcal{D}_t v^i(t, x) \geq f^i\left(t, x, v^i(t, x), \mathcal{D}_x v^i(t, x), \mathcal{D}_{xx} v^i(t, x), v\right) \quad for \ (t, x) \in D, \quad i \in S.$$

v. *Suppose finally that*

$$u(t, x) \leq v(t, x) \quad for \ (t, x) \in \overline{D}.$$

Under these assumptions

$$u(t, x) \leq v(t, x) \quad for \ (t, x) \in \overline{D}.$$

3.8 Strong Differential Inequalities for Infinite Systems

In the 1970s, theorems on weak differential inequalities were generalized to include infinite systems; however, theorems on strong differential inequalities have not been generalized in this manner. The difficulty lies in a proof of the existence of a *Nagumo point*. The Müller-Nagumo-Westphal Lemma on differential inequalities was proved and the notions of a *Nagumo point* and the *Nagumo method* of obtaining inequalities for solutions of parabolic inequalities were introduced.

In the theorems on strong differential inequalities, the fundamental assumption is some condition on monotonicity in y and s only, and in the theorems on weak differential inequalities, the fundamental assumptions are this monotonicity condition and the Lipschitz condition with respect to y and s. These theorems were proved in the Banach space $C_{\mathbb{N}}(\overline{D})$. It has unfortunately turned out, though, that no theorem on strong differential inequalities for infinite systems in the space $C_{\mathbb{N}}(\overline{D})$ can be proved. Precisely, the theorem fails to hold if the considered functions $u, v \in C_{\mathbb{N}}(\overline{D})$. A problem arises at the very beginning of the proof, when one attempts to use the Nagumo point method. Namely, if we deal with an infinitely countable system, that is if u, v are infinite sequences $u = \{u^k\}_{k\in\mathbb{N}}$ and $v = \{v^k\}_{k\in\mathbb{N}}$, then it may happen that

$$u^k(t, x) < v^k(t, x) \quad \text{for } (t, x) \in D, \quad 0 < t < \tilde{t} = \frac{1}{k}, \quad k \in \mathbb{N}.$$

Then, though, the intersection of the intervals $\left(0, \frac{1}{k}\right)$ for $k \in \mathbb{N}$ is empty and the inequality $u(t, x) < v(t, x)$ for $(t, x) \in D, 0 < t < t^*$ is not true; hence the proof fails. Therefore, we introduce the space $C_{\mathbb{N}}(\overline{D}; l^\infty)$ of all continuous functions from $\overline{D}$ into l^∞ equipped with the supremum norm from the space l^∞ and in this space we prove the theorem on strong differential inequalities. The gist of the idea thus consists in assuming the continuity of the functions considered and not only of the coordinates of these functions.

Theorem 3.8   *Let real functions $f^i(t, x, y, p, q, s), i \in \mathbb{N}$, be defined for $(t, x, y, p, q, s) \in \overline{D} \times l^\infty \times \mathbb{R}^m \times \mathcal{Q}_{m\times m} \times C_{\mathbb{N}}(\overline{D}; l^\infty)$, let the domain D have the property $\mathcal{P}$ and let $D_0 = \overline{D}$.*

Assume that:

i.   *the functions $f^i, i \in \mathbb{N}$, satisfy condition (**W+**) with respect to y^3 and condition ($\mathcal{M}$) with respect to s;*

ii.   *the functions $u, v \in C_{\mathbb{N}}^{\mathrm{reg}}(\overline{D}; l^\infty)$;*

iii.   *every function f^i is elliptic with respect to the function u in Szarski's sense;*

iv.   *the infinite systems of inequalities*

$$\mathcal{D}_t u^i(t, x) \le f^i\left(t, x, u(t, x), \mathcal{D}_x u^i(t, x), \mathcal{D}_{xx}^2 u^i(t, x), u\right), \tag{3.13}$$

and

$$\mathcal{D}_t v^i(t, x) > f^i\left(t, x, v(t, x), \mathcal{D}_x v^i(t, x), \mathcal{D}_{xx}^2 v^i(t, x), v\right), \quad i \in \mathbb{N}, \tag{3.14}$$

hold for $(t, x) \in D$.

v.   *Suppose finally that the initial inequality*

$$u(0, x) < v(0, x) \quad \text{for } x \in S_0 \tag{3.15}$$

and the boundary inequality

$$u(t, x) < v(t, x) \quad \text{for } (t, x) \in \sigma \tag{3.16}$$

hold true.

---

[3] Functions $f^i(t, x, y, p, q, s), i \in \mathbb{N}$, satisfy condition (**W+**) i.e., are quasi-increasing with respect to y if for arbitrary $y, \tilde{y} \in l^\infty$ there is

$$y \overset{(i)}{\le} \tilde{y} \Longrightarrow f^i(t, x, y, p, q, s) \le f^i(t, x, \tilde{y}, p, q, s)$$

for $(t, x) \in \overline{D}, p, q, s \in \mathbb{R}^m \times \mathcal{Q}_{m\times m} \times C_{\mathbb{N}}(\overline{D}; l^\infty)$.

If one of the inequalities (3.13) *or* (3.14) *is strict, then under the above assumptions there is*

$$u(t, x) < v(t, x) \quad for \ (t, x) \in \overline{D}. \tag{3.17}$$

Proof.   Notice that the proof of our theorem is similar to the proofs of Szarski's theorems on strong inequalities given in papers but it is based on the assumption that the functions $u, v \in \mathcal{C}_{\mathbb{N}}(\overline{D}; l^\infty)$.

Since the set of points $(0, x)$ such that $x \in S_0$ is compact, there is, by (3.15) and by the continuity of functions $u, v \in \mathcal{C}_{\mathbb{N}}(\overline{D}; l^\infty)$, a time $\tilde{t}, 0 < \tilde{t} < T$, such that (3.17) holds true in the intersection of $\overline{D}$ with the zone $0 \le t < \tilde{t}$. Let t^* be the least upper bound for such $\tilde{t}$, or $+\infty$ if there is no such bound. The assertion of the theorem is obviously equivalent to the equality $t^* = T$.

Suppose for the contrary that the conclusion is not true, i.e., $t^* < T$. Then by (ii) there is $u - v \in \mathcal{C}_{\mathbb{N}}(\overline{D}; l^\infty)$ and by the continuity of $u - v$ there would be

$$u(t, x) \le v(t, x) \quad for \ (t, x) \in \overline{D}, \quad 0 < t \le t^*. \tag{3.18}$$

This inequality means that

$$\overset{t^*}{u} \le v. \tag{3.19}$$

The domain D has property $\mathcal{P}$ and strong initial and boundary inequalities (3.15) and (3.16) hold; therefore, by the definition of t^* and the definition of order "<", there exists an index $i^* \in S$ and a point $x^* \in S_{t^*}$ such that

$$u^{i^*}(t^*, x^*) = v^{i^*}(t^*, x^*) \tag{3.20}$$

and (t^*, x^*) is an interior point of D.

The point (t^*, x^*) is called *a Nagumo point* and this method of proving is called the *Nagumo procedure (Nagumo method)*. The Nagumo procedure applies to parabolic equations containing a functional provided that the functional is monotone and of Volterra type.

From (3.18) and (3.20) it follows that the function

$$V(x) := u^{i^*}(t^*, x) - v^{i^*}(t^*, x)$$

as the function in $x = (x_1, \ldots, x_m)$ attains its maximum in S_{t^*} at $x = x^*$, i.e.,

$$\max_{x \in S_{t^*}} V(x) = \max_{x \in S_{t^*}} \left[u^{i^*}(t^*, x) - v^{i^*}(t^*, x) \right] = u^{i^*}(t^*, x^*) - v^{i^*}(t^*, x^*) = 0.$$

Since S_{t^*} is open and the function $V(x)$ is of class C^2 in S_{t^*} and attains its maximum at an interior point x^* of S_{t^*}, there is

$$\mathcal{D}_x u^{i^*}(t^*, x^*) = \mathcal{D}_x v^{i^*}(t^*, x^*) \tag{3.21}$$

and

$$\mathcal{D}^2_{xx} u^{i^*}(t^*, x^*) \leq \mathcal{D}^2_{xx} v^{i^*}(t^*, x^*). \tag{3.22}$$

Hence,

$$\sum_{j,k=1}^{m} \left[\mathcal{D}^2_{x_j x_k} u^{i^*}(t^*, x^*) - \mathcal{D}^2_{x_j x_k} v^{i^*}(t^*, x^*) \right] \xi_j \xi_k \leq 0 \tag{3.23}$$

for all $\xi = (\xi_1, \ldots, \xi_m) \in \mathbb{R}^m$.

From Assumptions (i)–(iv), (3.17)–(3.22), conditions (**V**) and (**W**), as well as interrelations between conditions (**V**) and (**W**) $\Leftrightarrow \mathcal{M}^4$ it follows that

$$
\begin{aligned}
\mathcal{D}_t u^{i^*}(t^*, x^*) &\leq f^{i^*}\left(t^*, x^*, u(t^*, x^*), \mathcal{D}_x u^{i^*}(t^*, x^*), \mathcal{D}^2_{xx} u^{i^*}(t^*, x^*), u\right) \\
&\leq f^{i^*}\left(t^*, x^*, u(t^*, x^*), \mathcal{D}_x u^{i^*}(t^*, x^*), \mathcal{D}^2_{xx} v^{i^*}(t^*, x^*), v\right) \\
&\leq f^{i^*}\left(t^*, x^*, v(t^*, x^*), \mathcal{D}_x v^{i^*}(t^*, x^*), \mathcal{D}^2_{xx} v^{i^*}(t^*, x^*), v\right) \\
&< \mathcal{D}_t v^{i^*}(t^*, x^*).
\end{aligned} \tag{3.24}
$$

On the other hand, the function

$$W(t) := u^{i^*}(t, x^*) - v^{i^*}(t, x^*),$$

as the function in one variable t, defined for t in some interval $(0, t^*)$ attains, by (3.18) and (3.20), its maximum at the right-hand extremity t^* of the interval $[0, t^*]$. Hence, there is

$$\mathcal{D}_t W(t^*) = \mathcal{D}_t u^{i^*}(t^*, x^*) - \mathcal{D}_t v^{i^*}(t^*, x^*) \geq 0, \tag{3.25}$$

which contradicts (3.24). Thus the theorem is proved. □

Now we formulate a theorem with another, more general boundary condition, to some extension corresponding to the first and the third classical Fourier problems. Let the functions

$$\alpha^i(t, x) \geq 0, \quad \beta^i(t, x) > 0 \quad \text{for } i \in \mathbb{N}, \tag{3.26}$$

be defined on σ and by Σ_{α^i} we denote the subset of σ on which $\alpha^i(t, x) \neq 0$. Suppose that for every point $(t, x) \in \Sigma_{\alpha^i}$ the direction $l^i = l^i(t, x)$ orthogonal to the t-axis and penetrating into the closed domain D is given.

The following theorem is true.

---

[4] Monotonicity condition ($\mathcal{M}$).

Theorem 3.9 *Under all assumptions of Theorem 3.8, with the exception of inequality (3.16), which is replaced with the following more general inequalities:*

$$\alpha^i(t, x)\frac{d[u^i - v^i]}{dl^i}(t, x) - \beta^i(t, x)[u^i(t, x) - v^i(t, x)] > 0 \tag{3.27}$$

for $(t, x) \in \Sigma_{\alpha^i}, i \in \mathbb{N}$, and

$$u^i(t, x) < v^i(t, x) \quad for \ (t, x) \in \sigma \setminus \Sigma_{\alpha^i}, \quad i \in \mathbb{N}, \tag{3.28}$$

where $\frac{du}{dl}$ is the directional derivative of z in the direction v on σ, inequality (3.17) is true, i.e., there is

$$u(t, x) < v(t, x) \quad for \ (t, x) \in \overline{D}.$$

The proof is quite similar to that of Theorem 3.8.

In the particular case of $\alpha = 0$ and $\beta = 1$, boundary inequality (3.27) is reduced to inequality (3.16).

Remark 3.2 Theorems 3.8 and 3.9 hold true for an arbitrary uncountable infinite system of inequalities (3.10). In this case we introduce the space $\mathcal{B}(S)$, where S is an arbitrary infinite set of indices, and the space $\mathcal{C}_S(\overline{D}; \mathcal{B}(S))$ as the space of all continuous functions from $\overline{D}$ into $\mathcal{B}(S)$, equipped with the supremum norm from the space $\mathcal{B}(S)$.

Remark 3.3 Theorems 3.8 and 3.9 hold true in unbounded domains for functions w which fulfill the growth condition $|w(t, x)| \leq M \exp\left(K |x|^2\right)$ in D, i.e., $w \in E_2(M, K; D)$.

CHAPTER 4

Monotone Iterative Methods

4.1 Method of Direct Iterations

We consider a cylindrical domain $D := (0, T] \times G$, where T is an arbitrary positive constant, G is an open and bounded domain in $\mathbb{R}^m$ whose boundary ∂G is an $(m-1)$-dimensional surface of a class $C^{2+\alpha}$ for some $0 < \alpha < 1$ and we use notations: $S_0 := \{(t, x) : t = 0, x \in \overline{G}\}$, the lateral surface $\sigma := (0, T] \times \partial G$, the parabolic boundary of domain D is $\Gamma := S_0 \cup \sigma$, and $\overline{D} := D \cup \Gamma$.

We assume that

$$f^i : \overline{D} \times \mathcal{B}(S) \times C_S(\overline{D}) \to \mathbb{R}, \quad (t, x, y, s) \mapsto f^i(t, x, y, s), \quad i \in S$$

are given functions and we consider an infinite system of equations with a homogeneous initial-boundary condition in $\overline{D}$, i.e., the problem

$$\begin{cases} \mathcal{F}^i[z^i](t, x) = f^i(t, x, z(t, x), z) & \text{for } (t, x) \in D, \\ z^i(t, x) = 0 & \text{for } (t, x) \in \Gamma, \quad \text{and } i \in S, \end{cases} \tag{4.1}$$

where the operators $\mathcal{F}^i$, $i \in S$, are diagonal and uniformly parabolic in $\overline{D}$.

The following theorem holds true.

Theorem 4.1 *Let assumptions A_0, (H_a), (H_f) hold, and conditions (L), $(\tilde{W})$, (V) hold in the set $\mathcal{K}$. If we define the successive terms of the approximation sequences $\{u_n\}$ and $\{v_n\}$ as regular solutions in $\overline{D}$ of the following infinite systems of equations:*

$$\mathcal{F}^i[u_n^i](t, x) = f^i\big(t, x, u_{n-1}(t, x), u_{n-1}\big) \quad \text{for } (t, x) \in D \quad \text{and } i \in S \tag{4.2}$$

and

$$\mathcal{F}^i[v_n^i](t, x) = f^i\big(t, x, v_{n-1}(t, x), v_{n-1}\big) \quad \text{for } (t, x) \in D \quad \text{and } i \in S \tag{4.3}$$

for $n = 1, 2, \ldots$ in D with homogeneous initial-boundary condition (2.6), and let an ordered pair of a lower u_0 and an upper solution v_0 (given by Assumption A_0), and such that $u_0 \leq v_0$ in $\overline{D}$, will be the pair of initial iterations in this iterative process, then

   i. *$\{u_n\}$, $\{v_n\}$ are well defined and $u_n, v_n \in C_S^{2+\alpha}(\overline{D})$ for $n = 1, 2, \ldots$;*

Mathematical Neuroscience. http://dx.doi.org/10.1016/B978-0-12-411468-5.00004-1

ii. *the inequalities*

$$u_0(t, x) \leq u_n(t, x) \leq u_{n+1}(t, x) \quad for \ n = 1, 2, \ldots, \tag{4.4}$$

hold for $(t, x) \in \overline{D}$ *and the functions* u_n *for* $n = 1, 2, \ldots,$ *are lower solutions of problem* (4.1) *in* $\overline{D}$, *and analogously*

$$v_{n+1}(t, x) \leq v_n(t, x) \leq v_0(t, x) \quad for \ n = 1, 2, \ldots, \tag{4.5}$$

hold for $(t, x) \in \overline{D}$ *and* v_n *for* $n = 1, 2, \ldots,$ *are upper solutions of problem* (4.1) *in* $\overline{D}$;

iii. *the inequalities*

$$u_n(t, x) \leq v_n(t, x) \quad n = 1, 2, \ldots, \tag{4.6}$$

hold for $(t, x) \in \overline{D}$;

iv. *the following estimate is true*

$$v_n^i(t, x) - u_n^i(t, x) \leq N_0 \frac{[(L_1 + L_2)t]^n}{n!} \quad i \in S, \quad n = 1, 2, \ldots \tag{4.7}$$

for $(t, x) \in \overline{D}$, *where* $N_0 = \|v_0 - u_0\|_0 = const < \infty$;

v. $\lim_{n \to \infty} \left[v_n^i(t, x) - u_n^i(t, x) \right] = 0$ *uniformly in* $\overline{D}$ *and* $i \in S$;

vi. *the function*

$$z = z(t, x) = \lim_{n \to \infty} u_n(t, x)$$

is the unique global (in time) regular solution of problem (2.29) *within the sector* $\langle u_0, v_0 \rangle$, *and* $z \in C_S^{2+\alpha}(\overline{D})$.

Before going into the proof of the theorem, we will introduce the fundamental operator $\mathcal{P}$, the nonlinear operator, and prove some lemmas. Since the proofs are straightforward and similar for the lower and upper solutions, we present the proof for the upper solution only.

Let $\beta = \beta(t, x)$ be a sufficiently regular function defined in $\overline{D}$. Denote by $\mathcal{P}$ the operator

$$\mathcal{P}: \beta \mapsto \mathcal{P}[\beta] = \gamma$$

generated by differential-functional problem in this way that the function $\gamma = \gamma(t, x)$ is the (supposedly unique) solution of the linear initial-boundary value problem

$$\begin{cases} \mathcal{F}^i[\gamma^i](t, x) = f^i(t, x, \beta(t, x), \beta) & for \ (t, x) \in D, \\ \gamma^i(t, x) = 0 & for \ (t, x) \in \Gamma \ and \ i \in S. \end{cases} \tag{4.8}$$

The operator $\mathcal{P}$ is the composition of the *nonlinear* operator $\mathbf{F} = \{F^i\}_{i \in S}$ generated by the functions $f^i(t, x, y, s), i \in S$, and defined for any sufficiently regular function β as follows:

$$\mathbf{F}: \beta \mapsto \mathbf{F}[\beta] = \delta,$$

where

$$\mathbf{F}^i[\beta](t, x) := f^i(t, x, \beta(t, x), \beta) := \delta^i(t, x) \quad \text{for } (t, x) \in D \quad \text{and} \quad i \in S, \tag{4.9}$$

and the operator

$$\mathcal{G}: \delta \mapsto \mathcal{G}[\delta] = \gamma,$$

where $\gamma = \gamma(t, x)$ is the (supposedly unique) solution of the linear initial-boundary value problem

$$\begin{cases} \mathcal{F}^i[\gamma^i](t, x) = \delta^i(t, x) & \text{for } (t, x) \in D, \\ \gamma^i(t, x) = 0 & \text{for } (t, x) \in \Gamma \quad \text{and} \quad i \in S. \end{cases} \tag{4.10}$$

Hence

$$\mathcal{P} = \mathcal{G} \circ \mathbf{F}.$$

The introduction of the nonlinear operator enables proofs of certain theorems to be simplified significantly and renders the proofs elegant. The examination of properties of the operator $\mathcal{P}$, including its monotonicity or compactness, is the founding idea of the theory of monotone iterative methods.

From the form of infinite system (4.8), precisely from the diagonal form of differential operators $\mathcal{F}^i$, $i \in S$, it follows that this weakly coupled system has the following fundamental property: the ith equation depends on the ith unknown function only. Roughly speaking, system (4.8) (and systems (4.2) and (4.3)) is a collection of several equations each of which is an equation in one unknown function only. Therefore, its solution is a collection of separate solutions of the scalar problems considered in suitably selected space.

Lemma 4.1 *If $\beta \in C_S^{0+\alpha}(\overline{D})$ and the function $f = \{f^i\}_{i \in S}$, generating the nonlinear operator* **F**, *satisfies conditions* (**H$_\mathbf{f}$**) *and* (**L**), *then*

$$\mathbf{F}: C_S^{0+\alpha}(\overline{D}) \to C_S^{0+\alpha}(\overline{D}) \quad \beta \mapsto \mathbf{F}[\beta] = \delta.$$

Proof. Because $\beta \in C_S^{0+\alpha}(\overline{D})$,

$$\left| \beta^i(t, x) - \beta^i(t', x') \right| \le H_\alpha^D(\beta) \left(|t - t'|^{\frac{\alpha}{2}} + |x - x'|^\alpha \right)$$

for (t, x), $(t', x') \in \overline{D}$ and all $i \in S$, where $H_\alpha^D(\beta) > 0$ is a constant independent of the index i.

From assumption (**H$_\mathbf{f}$**) and (**L**) it follows that

$$|\delta^i(t, x) - \delta^i(t', x')| = |\mathbf{F}^i[\beta](t, x) - \mathbf{F}^i[\beta](t', x')|$$
$$= |f^i(t, x, \beta(t, x), \beta) - f^i(t', x', \beta(t', x'), \beta)|$$

$$\leq |f^i(t, x, \beta(t, x), \beta) - f^i(t', x', \beta(t, x), \beta)|$$
$$+|f^i(t', x', \beta(t, x), \beta) - f^i(t', x', \beta(t', x'), \beta)|$$
$$\leq H_\alpha^D(f)\left(|t - t'|^{\frac{\alpha}{2}} + |x - x'|^\alpha\right) + L\,\|\beta(t, x) - \beta(t', x')\|_{B(S)}$$
$$\leq H_\alpha^D(f)\left(|t - t'|^{\frac{\alpha}{2}} + |x - x'|^\alpha\right) + LH_\alpha^D(\beta)\left(|t - t'|^{\frac{\alpha}{2}} + |x - x'|^\alpha\right)$$
$$\leq H_\alpha^D(\delta)\left(|t - t'|^{\frac{\alpha}{2}} + |x - x'|^\alpha\right)$$

for all $(t, x), (t', x') \in \overline{D}, i \in S$, where $H_\alpha^D(\delta) = H_\alpha^D(f) + LH_\alpha^D(\beta)$. Therefore, $\mathbf{F}[\beta] \in C_S^{0+\alpha}(\overline{D})$. This completes the proof. □

From Lemma 4.1 and Theorem A.1, the next lemma follows directly.

Lemma 4.2 *If $\delta \in C_S^{0+\alpha}(\overline{D})$, $\partial G \in C^{2+\alpha}$ and all the coefficients of the operators $\mathcal{L}^i, i \in S$, satisfy condition* ($\mathbf{H_a}$), *then problem (4.10) has the unique regular solution γ and $\gamma \in C_S^{2+\alpha}(\overline{D})$.*

Corollary 4.1 *From Lemmas 4.1 and 4.2 it follows that*

$$\mathcal{P} = \mathcal{G} \circ \mathbf{F} \colon C_S^{0+\alpha}(\overline{D}) \to C_S^{2+\alpha}(\overline{D}).$$

Lemma 4.3 *Let all the assumptions of Lemmas 4.1 and 4.2 hold, β be an upper solution and α be a lower solution of problem (4.1) in $\overline{D}, \alpha, \beta \in \langle u_0, v_0 \rangle$ and condition* ($\tilde{\mathbf{W}}$) *holds. Then*

$$\alpha(t, x) \leq \mathcal{P}[\beta](t, x) \leq \beta(t, x) \quad \text{for } (t, x) \in D \quad \text{and} \quad i \in S \tag{4.11}$$

and $\gamma = \mathcal{P}[\beta]$ is an upper solution of problem (4.1) in $\overline{D}$, and analogously

$$\alpha(t, x) \leq \mathcal{P}[\alpha](t, x) \leq \beta(t, x) \quad \text{for } (t, x) \in D \quad \text{and} \quad i \in S \tag{4.12}$$

and $\eta = \mathcal{P}[\alpha]$ is a lower solution of problem (4.1) in $\overline{D}$.

Proof. If β is an upper solution, then by virtue of (2.21) there is

$$\mathcal{F}^i[\beta^i](t, x) \geq f^i(t, x, \beta(t, x), \beta) \quad \text{for } (t, x) \in D \quad \text{and} \quad i \in S.$$

From the definition (4.8) of the operator $\mathcal{P}$ it follows that

$$\mathcal{F}^i[\gamma^i](t, x) = f^i(t, x, \beta(t, x), \beta) \quad \text{for } (t, x) \in D \quad \text{and} \quad i \in S.$$

Therefore

$$\mathcal{F}^i[\beta^i - \gamma^i](t, x) \geq 0 \quad \text{for } (t, x) \in D \quad \text{and} \quad i \in S$$

and

$$\beta(t, x) - \gamma(t, x) = 0 \quad \text{for } (t, x) \in \Gamma.$$

Hence, by Lemma 3.1 there is

$$\beta(t, x) - \gamma(t, x) \geq 0 \quad \text{for } (t, x) \in \overline{D}$$

so

$$\gamma(t, x) = \mathcal{P}[\beta](t, x) \leq \beta(t, x) \quad \text{for } (t, x) \in \overline{D}.$$

From (4.8) and conditions ($\tilde{\mathbf{W}}$) we obtain

$$\mathcal{F}^i[\gamma^i](t, x) - f^i(t, x, \gamma(t, x), \gamma) = f^i(t, x, \beta(t, x), \beta) - f^i(t, x, \gamma(t, x), \gamma) \geq 0$$

for $(t, x) \subset D, i \subset S$, and

$$\gamma(t, x) = 0 \quad \text{for } (t, x) \in \Gamma.$$

From Corollary 4.1 we infer that γ is a regular function and $\gamma \in C_S^{2+\alpha}(\overline{D})$, so it is an upper solution of problem (4.1) in $\overline{D}$ and from Remark 2.6 it follows that:

$$\alpha(t, x) \leq \gamma(t, x) = \mathcal{P}[\beta](t, x) \leq \beta(t, x) \quad \text{for } (t, x) \in \overline{D}.$$

This completes the proof. □

Lemma 4.4 *If all assumptions of Lemma 4.3 hold, $\alpha, \beta \in \langle u_0, v_0 \rangle$ and $\alpha(t, x) \leq \beta(t, x)$ in $\overline{D}$, then*

$$\mathcal{P}[\alpha](t, x) \leq \mathcal{P}[\beta](t, x) \quad \text{for } (t, x) \in \overline{D} \tag{4.13}$$

i.e., the operator $\mathcal{P}$ is isotone[1] in the sector $\langle u_0, v_0 \rangle$.

Proof. Let $\eta = \mathcal{P}[\alpha]$ and $\gamma = \mathcal{P}[\beta]$, then by (4.8)

$$\mathcal{F}^i[\eta^i](t, x) = f^i(t, x, \alpha(t, x), \alpha) \quad \text{for } (t, x) \in D \quad \text{and} \quad i \in S$$

and

$$\mathcal{F}^i[\gamma^i](t, x) = f^i(t, x, \beta(t, x), \beta) \quad \text{for } (t, x) \in D \quad \text{and} \quad i \in S$$

and

$$\eta(t, x) = \gamma(t, x) = 0 \quad \text{for } (t, x) \in \Gamma.$$

By ($\tilde{\mathbf{W}}$), there is

$$\mathcal{F}^i[\eta^i - \gamma^i](t, x) = f^i(t, x, \beta(t, x), \beta) - f^i(t, x, \alpha(t, x), \alpha) \geq 0$$
$$\text{for } (t, x) \in D \quad \text{and} \quad i \in S$$

---

[1] Let $\mathcal{X}$ and $\mathcal{Y}$ be partially ordered sets with the ordering given by cones $K_{\mathcal{X}}$ and $K_{\mathcal{Y}}$ and denoted by "$\leq$" in each set. A transformation $\mathcal{T}: \mathcal{X} \to \mathcal{Y}$ is called *isotone* (monotone increasing) if for each $u, v \in \mathcal{X}, u \leq v$ implies $\mathcal{T}[u] \leq \mathcal{T}[v]$ and *strictly isotone* if $u < v$ implies $\mathcal{T}[u] < \mathcal{T}[v]$.

and
$$\gamma(t, x) - \eta(t, x) = 0 \quad \text{on } \Gamma.$$

By Lemma 3.1, there is

$$\gamma(t, x) \geq \eta(t, x) \quad \text{in } \overline{D}, \quad \text{i.e.,} \quad P[\alpha](t, x) \leq P[\beta](t, x) \quad \text{in } \overline{D}.$$

This means that the operator P is isotone in the sector $\langle u_0, v_0 \rangle$. This completes the proof.   □

From Corollary 4.1, Lemmas 4.3 and 4.4, the next corollary follows.

Corollary 4.2 *If all the assumptions of Lemma 4.3 hold and if α and β are a lower and an upper solution of problem (4.1) in $\overline{D}$, respectively, $\alpha, \beta \in \langle u_0, v_0 \rangle$, and $\alpha(t, x) \leq \beta(t, x)$ then*

$$\alpha(t, x) \leq P[\alpha](t, x) \leq P[\beta](t, x) \leq \beta(t, x) \quad \text{for } (t, x) \in \overline{D}. \tag{4.14}$$

This means that $P[\langle \alpha, \beta \rangle] \subset \langle \alpha, \beta \rangle$.

Proof of Theorem 4.1. Starting from the lower solution u_0 and the upper solution v_0, we define by induction two sequences of functions $\{u_n\}$ and $\{v_n\}$ as regular solutions of systems (4.2) and (4.3). Therefore we have

$$u_n = P[u_{n-1}], \quad v_n = P[v_{n-1}] \quad \text{for } n = 1, 2, \ldots.$$

From Lemmas 4.1, 4.2, and 4.4 it follows that u_n and v_n, for $n = 1, 2, \ldots$, are well defined and are the lower and the upper solutions of problem (4.2) and (4.3) in $\overline{D}$, respectively.

Using mathematical induction, from Lemma 4.3, we obtain

$$u_n(t, x) \leq P[u_n](t, x) = u_{n+1}(t, x), \quad n = 1, 2, \ldots$$

and
$$v_{n+1}(t, x) = P[v_n](t, x) \leq v_n(t, x) \quad \text{for } (t, x) \in \overline{D} \quad \text{and} \quad n = 1, 2, \ldots.$$

Therefore, inequalities (4.4) and (4.5) hold.

Analogously, using mathematical induction, from Lemma 4.4 and Assumption A_0 we also obtain inequalities (4.6).

Using mathematical induction, we will prove the inequalities

$$0 \leq v_n^i(t, x) - u_n^i(t, x) := w_n^i(t, x) \leq N_0 \frac{[(L_1 + L_2)t]^n}{n!},$$
$$\text{for } (t, x) \in \overline{D} \quad \text{and} \quad i \in S, \tag{4.15}$$

$n = 1, 2, \ldots.$

It is obvious that inequality (4.15) holds for w_0. Let inequality (4.15) hold for w_n. The functions $f^i(t, x, y, s)$, $i \in S$, fulfill Lipschitz condition (**L**) with respect to y and s, and condition (**V**). Therefore , using the interrelation between conditions (**V**) and (**L**) $\Longleftrightarrow$ (**L***) (see Section 2.7), f^i fulfills condition (**L***). By (4.2), (4.3), and (4.15), there is

$$\mathcal{F}^i \left[w^i_{n+1} \right] (t, x) = f^i(t, x, v_n(t, x), v_n) - f^i(t, x, u_n(t, x), u_n)$$
$$\leq L_1 \| w_n(t, x) \|_{\mathcal{B}(S)} + L_2 \| w_n \|_{0,t} .$$

By the definition of the norm $\| \cdot \|_{0,t}$ in the space $C_S(\overline{D})$ and by inequality (4.15),

$$\| w_n \|_{0,t} \leq \frac{[(L_1 + L_2)t]^n}{n!} .$$

So we finally obtain

$$\mathcal{F}^i \left[w^i_{n+1} \right] (t, x) \leq N_0 \frac{(L_1 + L_2)^{n+1} t^n}{n!} \quad \text{for } (t, x) \in D \quad \text{and} \quad i \in S \tag{4.16}$$

and

$$w_{n+1}(t, x) = 0 \quad \text{for } (t, x) \in \Gamma. \tag{4.17}$$

Let us consider the comparison system of equations

$$\mathcal{F}^i \left[M^i_{n+1} \right] (t, x) - N_0 \frac{(L_1 + L_2)^{n+1} t^n}{n!} \quad \text{for } (t, x) \in D \quad \text{and} \quad i \in S \tag{4.18}$$

with the initial-boundary condition

$$M_{n+1}(t, x) \geq 0 \quad \text{for } (t, x) \in \Gamma \tag{4.19}$$

It is obvious that the functions

$$M^i_{n+1}(t, x) = N_0 \frac{[(L_1 + L_2)t]^{n+1}}{(n+1)!} \quad \text{for } (t, x) \in \overline{D} \quad \text{and} \quad i \in S$$

are the regular solutions of comparison problem (4.18) and (4.19) in $\overline{D}$.

Applying the Szarski theorem on weak differential inequalities (Theorem 3.1) to problems (4.16), (4.17) and (4.18), (4.19), we obtain

$$w^i_{n+1}(t, x) \leq M^i_{n+1}(t, x) = N_0 \frac{[(L_1 + L_2)t]^{n+1}}{(n+1)!} \quad \text{for } (t, x) \in \overline{D} \quad \text{and} \quad i \in S$$

so the induction step is proved and so estimate (4.7) is true.

As a direct conclusion from formula (4.15) we obtain

$$\lim_{n \to \infty} \left[v_n^i(t, x) - u_n^i(t, x) \right] = 0 \quad \text{uniformly in } \overline{D} \quad \text{for all } i \in S. \tag{4.20}$$

The sequences of functions $\{u_n(t, x)\}$ and $\{v_n(t, x)\}$ are monotonous and bounded, and (4.20) holds, so there exists a continuous function $U = U(t, x)$ in $\overline{D}$ such that

$$\lim_{n \to \infty} u_n^i(t, x) = U^i(t, x) \quad \text{and} \quad \lim_{n \to \infty} v_n^i(t, x) = U^i(t, x) \tag{4.21}$$

uniformly in $\overline{D}$, $i \in S$.

Since functions f^i $(i \in S)$ are monotonous (condition $\tilde{\mathbf{W}}$), from (4.4) it follows that the functions $f^i(t, x, u_{n-1}(t, x), u_{n-1})$, $i \in S$, are uniformly bounded in D with respect to n. Hence by Theorem A.1 we conclude that all the functions $u_n \in C_S^{2+\alpha}(\overline{D})$ for $n = 1, 2, \ldots$, satisfy the Hölder condition with a constant independent of n. Hence $U \in C_S^{0+\alpha}(\overline{D})$.

If we now consider the system of equations

$$\mathcal{F}^i[z^i](t, x) = f^i\big(t, x, U(t, x), U\big) := \mathbf{F}^i[U](t, x) \quad \text{for } (t, x) \in D \quad i \in S \tag{4.22}$$

with initial-boundary condition (2.6), then by Lemma 4.1 there is $\mathbf{F}^i[U] \in C_S^{0+\alpha}(\overline{D})$. Therefore, by virtue of Lemma 4.2 this problem has the unique regular solution z, and $z \in C_S^{2+\alpha}(\overline{D})$.

Let us now consider systems (4.2) and (4.22) together. Let us apply, to these systems, Szarski's theorem on the continuous dependence of the solution of the first problem on the initial-boundary values and on the right-hand sides of systems. Since the functions $f^i, i \in S$, satisfy the Lipschitz condition $(\mathbf{L})$, by (4.21), there is

$$\lim_{n \to \infty} f^i(t, x, u_n(t, x), u_n) = f^i\big(t, x, U(t, x), U\big) \quad \text{uniformly in } \overline{D}, \quad i \in S.$$

Hence

$$\lim_{n \to \infty} u_n^i(t, x) = z^i(t, x) \quad \text{for } i \in S. \tag{4.23}$$

By virtue of (4.21) and (4.23),

$$z = z(t, x) = U(t, x) \quad \text{for } (t, x) \in \overline{D}$$

is the regular solution of problem (4.1) in $\overline{D}$ and $z \in C_S^{2+\alpha}(\overline{D})$.

The uniqueness of the solution follows directly from Szarski's uniqueness criterion (Theorem 3.6). It also follows directly from inequality (4.15). Because the assumptions of our theorem hold in the set $\mathcal{K}$ only, the uniqueness of a solution is ensured only with respect to the given upper and lower solutions, and it does not rule out the existence of other solutions outside the sector $\langle u_0, v_0 \rangle$.

Thus the theorem is proved. $\square$

Remark 4.1 In the case of estimate (4.15) we say that the sequences of successive approximations $\{u_n\}$ and $\{v_n\}$ defined by (4.2) and (4.3) converge to the searched solution z with the power speed.

Remark 4.2 The convergence of the method of direct iterations may also be derived from the comparison theorem. The comparison theorem allows us to prove that the sequences of direct iterations $\{u_n\}$ and $\{v_n\}$ are Cauchy sequences.

From Theorem 4.1, as a particular case there follows a theorem on the existence and uniqueness of a solution of the Fourier first initial-boundary value problem for the infinite system of equations. This theorem is a generalization of the well-known Kusano theorem to the case of the infinite system.

Theorem 4.2 *Let us consider the Fourier first initial-boundary value problem for the infinite system of equations of the form*

$$\begin{cases} \mathcal{F}^i[z^i](t, x) = \tilde{f}^i(t, x, z(t, x)) & \text{for } (t, x) \in D, \\ z^i(t, x) = 0 & \text{for } (t, x) \in \Gamma \quad \text{and} \quad i \in S. \end{cases} \tag{4.24}$$

Let the following assumptions be satisfied:

i.   *the operators $\mathcal{F}^i$, $i \in S$, are uniformly parabolic in $\overline{D}$;*
ii. *Assumption A_0 holds;*
iii. *all the coefficients of the operators $\mathcal{F}^i$, $i \in S$, fulfill condition $(\mathcal{H}_a)$;*
iv. *the functions $\tilde{f}^i(t, x, y)$, $i \in S$, are defined for $(t, x, y) \in \overline{D} \times B(S)$, satisfy assumption $(\mathbf{H_f})$, Lipschitz condition $(\mathbf{L})$, and condition $(\mathbf{W})$ with respect to y;*
v.   *the above assumptions hold locally in the sector $\langle u_0, v_0 \rangle$ formed by u_0 and v_0.*

Under these assumptions, problem (4.24) possesses the unique global regular solution z within the sector $\langle u_0, v_0 \rangle$, and $z \in C_S^{2+\alpha}(\overline{D})$.

4.2 Chaplygin Method

To solve problem (4.1), we now apply another monotone iterative method, namely the Chaplygin method, in which we use the linearization with respect to the nonfunctional argument y only.

We assume that the functions $f^i(t, x, y, s)$, $i \in S$, satisfy conditions $(\mathbf{H_f})$, $(\mathbf{L})$, $(\tilde{\mathbf{W}})$, and $(\mathbf{V})$ with respect to y and s in the set $\mathcal{K}$.

We additionally assume that each function $f^i(t, x, y, s)$, $i \in S$, has the continuous derivatives $\mathcal{D}_{y^i} f^i := \frac{\partial f^i}{\partial y^i} := f_{y^i}^i(t, x, y, s)$, $i \in S$, which satisfy the following conditions in the set $\mathcal{K}$:

Assumption H_p. Condition (H_f).

Assumption L_p. The Lipschitz condition with respect to y and s.

Assumption $\tilde{W}_p$. They are increasing with respect to y and s.

Theorem 4.3 *Let assumptions A_0, (H_a), (H_f), (H_p) hold, and conditions $(\tilde{W})$, $(\tilde{W}_p)$, (L), (L_p), (V) hold in the set $\mathcal{K}$. Let us assume that the successive terms of approximation sequences $\{\hat{u}_n\}$ and $\{\hat{v}_n\}$ are defined as regular solutions in $\overline{D}$ of the following infinite systems of linear equations:*

$$\mathcal{F}^i\left[\hat{u}_n^i\right](t, x) = f^i(t, x, \hat{u}_{n-1}(t, x), \hat{u}_{n-1})$$
$$+ f_{y^i}^i(t, x, \hat{u}_{n-1}(t, x), \hat{u}_{n-1}) \cdot \left[\hat{u}_n^i(t, x) - \hat{u}_{n-1}^i(t, x)\right], \quad i \in S \quad (4.25)$$

and

$$\mathcal{F}^i\left[\hat{v}_n^i\right](t, x) = f^i(t, x, \hat{v}_{n-1}(t, x), \hat{v}_{n-1}) + f_{y^i}^i(t, x, \hat{v}_{n-1}(t, x), \hat{v}_{n-1})$$
$$\cdot \left[\hat{v}_n^i(t, x) - \hat{v}_{n-1}^i(t, x)\right] \quad \text{for } (t, x) \in D, \quad i \in S \quad (4.26)$$

for $n = 1, 2, \ldots$, in D with homogeneous initial-boundary condition (2.6). Let us use $\hat{u}_0$ for u_0 and $\hat{v}_0$ for v_0 (the initial iteration in this iterative process).

Then

i. *$\{\hat{u}_n\}$, $\{\hat{v}_n\}$ are well defined and $\hat{u}_n, \hat{v}_n \in C_S^{2+\alpha}(\overline{D})$ for $n = 1, 2, \ldots$;*

ii. *the inequalities*

$$u_0(t, x) \le \hat{u}_n(t, x) \le \hat{u}_{n+1}(t, x) \le \hat{v}_{n+1}(t, x) \le \hat{v}_n(t, x) \le v_0(t, x) \quad (4.27)$$

hold for $(t, x) \in \overline{D}, n = 1, 2, \ldots$, and the functions $\hat{u}_n$ and $\hat{v}_n$ for $n = 1, 2, \ldots$, are lower and upper solutions of problem (4.1) in $\overline{D}$, respectively;

iii. *the following inequalities:*

$$u_n(t, x) \le \hat{u}_n(t, x) \le \hat{v}_n(t, x) \le v_n(t, x) \quad (4.28)$$

hold for $(t, x) \in \overline{D}$ and $n = 1, 2, \ldots$, where the sequences $\{u_n\}$ and $\{v_n\}$ are defined by (4.2) and (4.3);

iv. *the following estimate:*

$$\hat{v}_n^i(t, x) - \hat{u}_n^i(t, x) \le N_0 \frac{[(L_1 + L_2)t]^n}{n!}, \quad \text{holds for } (t, x) \in \overline{D},$$
$$i \in S, \quad n = 1, 2, \ldots, \quad (4.29)$$

where $N_0 = \|v_0 - u_0\|_0 = const < \infty$;

v.

$$\lim_{n\to\infty}\left[\hat{v}_n^i(t,x)-\hat{u}_n^i(t,x)\right]=0 \quad \text{uniformly in } \overline{D}, \quad i\in S;$$

vi. *the function*

$$z=z(t,x)=\lim_{n\to\infty}\hat{u}_n(t,x)$$

is the unique global regular solution of problem (4.1) within the sector $\langle u_0,v_0\rangle$, *and* $z\in C_S^{2+\alpha}(\overline{D})$.

Before proving the theorem, we introduce the nonlinear operators and prove some lemmas.

Let $\beta\in C_S(\overline{D})$ be a sufficiently regular function. Let $\hat{\mathcal{P}}$ denote the operator

$$\hat{\mathcal{P}}:\beta\mapsto\hat{\mathcal{P}}[\beta]=\gamma,$$

where γ is the (supposedly unique) solution of the following problem:

$$\begin{cases} \mathcal{F}^i[\gamma^i](t,x)+f_{y^i}^i(t,x,\beta(t,x),\beta)\cdot\gamma^i(t,x) \\ = f^i(t,x,\beta(t,x),\beta)+f_{y^i}^i(t,x,\beta(t,x),\beta)\cdot\beta^i(t,x) \quad \text{for } (t,x)\in D, \\ \gamma^i(t,x)=0 \quad \text{for } (t,x)\in\Gamma \quad \text{and} \quad i\in S. \end{cases} \quad (4.30)$$

Observe that system (4.30) has the following property: the ith equation depends on the ith unknown function only. Therefore, its solution is a collection of separate solutions of these scalar problems for linear equations in a suitable space.

It is convenient to define two nonlinear operators related to the functions $f^i(t,x,y,s)$ and $f_{y^i}^i(t,x,y,s)$, $i\in S$, to examine them separately. They are: the operator $\mathbf{C}=\{\mathbf{C}^i\}_{i\in S}$

$$\mathbf{C}:\beta\mapsto\mathbf{C}[\beta]=\eta,$$

where

$$\mathbf{C}^i[\beta](t,x):=f_{y^i}^i(t,x,\beta(t,x),\beta)=\eta^i(t,x) \quad (4.31)$$

for $(t,x)\in D$, $i\in S$ and the operator $\hat{\mathbf{F}}=\{\hat{\mathbf{F}}^i\}_{i\in S}$

$$\hat{\mathbf{F}}:\beta\mapsto\hat{\mathbf{F}}[\beta]=\delta,$$

where

$$\begin{aligned} \hat{\mathbf{F}}^i[\beta](t,x) &:= f^i(t,x,\beta(t,x),\beta)+f_{y^i}^i(t,x,\beta(t,x),\beta)\cdot\beta^i(t,x) \\ &= f^i(t,x,\beta(t,x),\beta)+\mathbf{C}^i[\beta](t,x)\cdot\beta^i(t,x):=\delta^i(t,x), \quad (4.32) \\ &\qquad\qquad \text{for } (t,x)\in D \quad \text{and} \quad i\in S. \end{aligned}$$

One may use the notation just introduced to write problem (4.30) in a simpler way.

The operator $\hat{P}$ is the composition of the nonlinear operators $\hat{\mathbf{F}}$ and $\mathbf{C}$ with the operator

$$\hat{\mathcal{G}}: \delta \mapsto \mathcal{G}[\delta] = \gamma,$$

where γ is the (supposedly unique) solution of the linear problem

$$\begin{cases} \mathcal{F}^i[\gamma^i](t, x) + \mathbf{C}^i[\beta](t, x) \cdot \gamma^i(t, x) = \hat{\mathbf{F}}^i[\beta](t, x) & \text{for } (t, x) \in D, \\ \gamma^i(t, x) = 0 & \text{for } (t, x) \in \Gamma \text{ and } i \in S. \end{cases} \tag{4.33}$$

Hence

$$\hat{P} = \hat{\mathcal{G}} \circ (\hat{\mathbf{F}}, \mathbf{C}).$$

The following lemmas hold, too.

Lemma 4.5   *If $\beta \in C_S^{0+\alpha}(\overline{D})$ and the functions $f_{y^i} = \{f_{y^i}^i\}$ generating the nonlinear operator $\mathbf{C}$ satisfy conditions $(\mathbf{H_p})$ and $(\mathbf{L_p})$, then*

$$\mathbf{C}[\beta] = \eta \in C_S^{0+\alpha}(\overline{D}).$$

If $\beta \in C_S^{0+\alpha}(\overline{D})$, the functions $f = \{f^i\}_{i \in S}$ and $f_{y^i} = \{f_{y^i}^i\}_{i \in S}$ generating the nonlinear operator $\hat{\mathbf{F}}$ satisfy conditions $(\mathbf{H_f})$, $(\mathbf{L})$, $(\mathbf{H_p})$, and $(\mathbf{L_p})$, then

$$\hat{\mathbf{F}}[\beta] = \delta \in C_S^{0+\alpha}(\overline{D}).$$

Proof.   It runs analogously to the proof of Lemma 4.1. This completes the proof.   □

From Theorem 3.3 and Lemma 4.5, one may derive the following statements.

Lemma 4.6   *If $\beta \in C_S^{0+\alpha}(\overline{D})$ and all the coefficients of the operators $\mathcal{L}^i$, $i \in S$ satisfy assumption $(\mathbf{H_a})$, then problem (4.33) has exactly one regular solution, $\gamma \in C_S^{2+\alpha}(\overline{D})$.*

Corollary 4.3   *It follows from Lemmas 4.5 and 4.6 that*

$$\hat{P}: C_S^{0+\alpha}(\overline{D}) \to C_S^{2+\alpha}(\overline{D}).$$

Lemma 4.7   *Let all the assumptions of Lemmas 4.5 and 4.6 hold, β be an upper solution and α be a lower solution of problem (4.1) in $\overline{D}$, and conditions $(\tilde{\mathbf{W}})$, $(\tilde{\mathbf{W}}_p)$ hold. Then*

$$\hat{P}[\beta](t, x) \le \beta(t, x) \quad \text{for } (t, x) \in \overline{D} \tag{4.34}$$

and $\gamma = \hat{P}[\beta]$ is an upper solution of problem (4.1) for $(t, x) \in \overline{D}$, and analogously

$$\hat{P}[\alpha](t, x) \ge \alpha(t, x) \quad \text{for } (t, x) \in \overline{D} \tag{4.35}$$

and $\eta = \hat{P}[\alpha]$ is a lower solution of problem (4.1) in $\overline{D}$.

Proof. If β is an upper solution, then due to (2.21) and the notation introduced, there is

$$\mathcal{F}^i[\beta^i](t, x) \geq f^i(t, x, \beta(t, x), \beta), \quad i \in S, \quad \text{for } (t, x) \in D$$

thus β satisfies the following system of inequalities:

$$\mathcal{F}^i[\tilde{z}^i](t, x) \geq f^i(t, x, \beta(t, x), \beta) + f^i_{y^i}(t, x, \beta(t, x), \beta) \cdot \left[\tilde{z}^i(t, x) - \beta^i(t, x)\right], \quad (4.36)$$

$i \in S$, for $(t, x) \in D$ and initial-boundary condition (2.6).

From definition (4.30) of the operator $\hat{\mathcal{P}}$ it follows that the function γ is a solution of the system of equations

$$\mathcal{F}^i[\tilde{\tilde{z}}^i](t, x) = f^i(t, x, \beta(t, x), \beta) + f^i_{y^i}(t, x, \beta(t, x), \beta) \cdot \left[\tilde{\tilde{z}}^i(t, x) - \beta^i(t, x)\right] \quad (4.37)$$

$i \in S$, for $(t, x) \in D$, with condition (2.6).

Applying Theorem 4.2 to systems (4.36) and (4.37), we obtain

$$\tilde{\tilde{z}}^i(t, x) \leq \tilde{z}^i(t, x), \quad i \in S, \quad \text{for } (t, x) \in \overline{D}$$

so,

$$\gamma(t, x) = \hat{\mathcal{P}}[\beta](t, x) \leq \beta(t, x) \quad \text{for } (t, x) \in \overline{D}.$$

Since the functions $f^i_{y^i}(t, x, y, s)$ are increasing in variable y^i (condition $(\tilde{\mathbf{W}}_{\mathbf{p}})$), then the functions $f^i(t, x, y, s)$ are convex in y^i.

Now (4.30), (4.34), and condition $(\tilde{\mathbf{W}})$ give

$$\mathcal{F}^i[\gamma^i](t, x) - f^i(t, x, \beta(t, x), \beta) + f^i_{y^i}(t, x, \beta(t, x), \beta) \cdot [\gamma^i(t, x) - \beta^i(t, x)]$$
$$\geq f^i(t, x, \gamma(t, x), \beta) \geq f^i(t, x, \gamma(t, x), \gamma), \quad i \in S, \quad \text{for } (t, x) \in \overline{D}$$

and

$$\gamma(t, x) = 0 \quad \text{on for } (t, x) \in \Gamma.$$

Thus from (2.21) and Corollary 4.2 it follows that the function γ is an upper solution of problem (4.1) for $(t, x) \in \overline{D}$. This completes the proof. $\square$

Proof of Theorem 4.3. One may prove statements (i)–(iii) of the theorem using the induction argument. Using Assumption A_0 it is easy to see that those statements are immediate consequences of Lemmas 4.5–4.7 and Corollary 4.3.

Indeed, due to the definition of $\hat{\mathbf{P}}$, (4.30), (4.25), and (4.26), we obtain

$$\hat{u}_n = \hat{\mathcal{P}}[\hat{u}_{n-1}], \quad \hat{v}_n = \hat{\mathcal{P}}[\hat{v}_{n-1}] \quad \text{for } n = 1, 2, \ldots$$

which means that the sequences $\{\hat{u}_n\}$ and $\{\hat{v}_n\}$ are well defined.

Statement (iv) follows immediately from the comparison of system (4.25) and (4.26) with system (4.2) and (4.3) defining the sequences $\{u_n\}$ and $\{v_n\}$.

Indeed, due to (4.30) and condition $(\tilde{\mathbf{W}}_{\mathbf{p}})$, there is

$$
\begin{aligned}
\mathcal{F}^i[\hat{u}_1^i](t, x) &- \mathcal{F}^i[u_1^i](t, x) \\
&= f^i(t, x, u_0(t, x), u_0) + f^i_{y^i}(t, x, u_0(t, x), u_0) \cdot \left[\hat{u}_1^i(t, x) - u_0^i(t, x)\right] \\
&\quad - f^i(t, x, u_0(t, x), u_0) \\
&= f^i_{y^i}(t, x, u_0(t, x), u_0) \cdot \left[\hat{u}_1^i(t, x) - u_0^i(t, x)\right] \geq 0 \\
&\qquad \text{for } (t, x) \in D \quad \text{and} \quad i \in S
\end{aligned}
$$

and

$$
\hat{u}_1(t, x) - u_1(t, x) = 0 \quad \text{for } (t, x) \in \Gamma.
$$

Consequently, by Lemma 3.1

$$
\hat{u}_1(t, x) \geq u_1(t, x) \quad \text{for } (t, x) \in \overline{D}.
$$

By mathematical induction, we obtain

$$
u_n(t, x) \leq \hat{u}_n(t, x) \leq \hat{v}_n(t, x) \leq v_n(t, x) \quad \text{for } (t, x) \in \overline{D} \quad \text{and} \quad n = 1, 2, \dots.
$$

The other statements, (v) and (vi), follow immediately from inequality (4.29) and Theorem 4.1. Thus the theorem is proved. $\square$

4.3 Certain Variants of the Chaplygin Method

Now we present the next two monotone iterative methods being certain variants of the Chaplygin method, applicable to problem (4.1), whose right-hand sides of equations are quasi-increasing functions; more precisely, they meet condition $(\mathbf{K})$. The last of these methods consists of adding appropriate linear terms $k^i(t, x)z^i$ including the unknown functions $z^i = z^i(t, x)$ to both sides of (2.2), in order to render the new sides monotonous. Thus we obtain the system of equations

$$
\mathcal{F}^i[z^i](t, x) + k^i(t, x)z^i(t, x) = f^i(t, x, z(t, x), z) + k^i(t, x)z^i(t, x) \quad \text{for } i \in S,
$$

whose right-hand sides are increasing with respect to the variable y^i for each i, $i \in S$.

Setting

$$
\mathcal{F}^i_\kappa := \mathcal{D}_t - \mathcal{L}^i + k^i \mathcal{I} \tag{4.38}
$$

we obtain a system of the following form:

$$\mathcal{F}_k^i[z^i](t, x) = f^i(t, x, z(t, x), z) + k^i(t, x)z^i(t, x)$$
$$\text{for } (t, x) \in D \quad \text{and} \quad i \in S \tag{4.39}$$

to which monotone iterative methods, including in particular the method of direct iteration, are applicable.

Theorem 4.4 *Let assumptions* A_0*,* $(\mathbf{H_a})$*,* $(\mathbf{H_f})$ *hold, and conditions* $(\tilde{\mathbf{W}})$*,* $(\mathbf{L})$*,* $(\mathbf{K})$*,* $(\mathbf{V})$ *hold in the set* $\mathcal{K}$*. Let the successive terms of the approximation sequences* $\{\overset{*}{u}_n\}$ *and* $\{\overset{*}{v}_n\}$ *be defined as regular solutions in* $\overline{D}$ *of the following infinite systems of equations:*

$$\mathcal{F}_k^i\left[\overset{*i}{u}_n\right](t, x) - f^i(t, x, \overset{*}{u}_{n-1}(t, x), \overset{\downarrow}{u}_{n-1}) + k^i(t, x)\overset{*i}{u}_{n-1}(t, x) \tag{4.40}$$
$$\text{for } (t, x) \in D \quad \text{and} \quad i \in S$$

and

$$\mathcal{F}_k^i\left[\overset{*}{v}_n^i\right](t, x) = f^i(t, x, \overset{*}{v}_{n-1}(t, x), \overset{*}{v}_{n-1}) + k^i(t, x)\overset{*i}{v}_{n-1}(t, x) \tag{4.41}$$
$$\text{for } (t, x) \in D \quad \text{and} \quad i \in S$$

for $n = 1, 2, \ldots$ *with homogeneous initial-boundary condition (2.6) where the function* $k^i = k^i(t, x)$*,* $i \in S$*, fulfill the assumption* $(\mathbf{H_a})$ *and let* $\overset{*}{u}_0 = u_0$*,* $\overset{*}{v}_0 = v_0$*.*

Then the statements of Theorem 4.3 hold, thus there exists the unique time-global regular solution z *of problem (4.1) within the sector* $\langle u_0, v_0 \rangle$*, and* $z \in C_S^{2+\alpha}(\overline{D})$*.*

Proof.   After some small changes in technical details, the proof of this theorem is identical to those of previous theorems, so we may omit it here.

If condition $(\mathbf{K}_\kappa)$ holds, then we will define the approximation sequences in the following way.

Theorem 4.5 *Let assumptions* A_0*,* $(\mathbf{H_a})$*,* $(\mathbf{H_f})$ *hold, and conditions* $(\mathbf{K}_\kappa)$*,* $(\tilde{\mathbf{W}})$*,* $(\mathbf{L})$*,* $(\mathbf{V})$ *hold in the set* $\mathcal{K}$*. Let the successive terms of the approximation sequences* $\{\overset{\circ}{u}_n\}$ *and* $\{\overset{\circ}{v}_n\}$ *be defined as regular solutions in* $\overline{D}$ *of the following infinite systems of equations:*

$$\mathcal{F}_\kappa^i\left[\overset{\circ i}{u}_n\right](t, x) = f^i(t, x, \overset{\circ}{u}_{n-1}(t, x), \overset{\circ}{u}_{n-1}) + \kappa\overset{\circ i}{u}_{n-1}(t, x) \tag{4.42}$$
$$\text{for } (t, x) \in D \quad \text{and} \quad i \in S$$

and

$$\mathcal{F}_\kappa^i\left[\overset{\circ i}{v}_n\right](t, x) = f^i(t, x, \overset{\circ}{v}_{n-1}(t, x), \overset{\circ}{v}_{n-1}) + \kappa\overset{\circ i}{v}_{n-1}(t, x) \tag{4.43}$$
$$\text{for } (t, x) \in D \quad \text{and} \quad i \in S$$

for n = 1, 2, ... for (t, x) ∈ D with the condition (2.6) and let $\mathring{u}_0 = u_0$, $\mathring{v}_0 = v_0$. Then there exists the unique global regular solution z of problem (4.1)

$$z = z(t, x) = \lim_{n \to \infty} \mathring{u}_n(t, x) = \lim_{n \to \infty} \mathring{v}_n(t, x)$$

within the sector $\langle u_0, v_0 \rangle$*, and* $z \in C_S^{2+\alpha}(\overline{D})$*.*

Remark 4.3 Let u_0 be a lower and v_0 be an upper solution on the initial-boundary value problem for equation

$$\begin{cases} (\mathcal{D}_t - \mathcal{L})[z](t, x) = \overline{\mathbf{F}}[z](t, x) := \overline{f}(t, x, z(t, x)) & \text{for } (t, x) \in D, \\ z(t, x) = 0 \quad \text{on } \Gamma \end{cases} \tag{4.44}$$

and suppose $u_0 \le v_0$ in $\overline{D}$.

Let

$$\alpha := \min_{\overline{D}} u_0(t, x) \quad \text{and} \quad \beta := \max_{\overline{D}} v_0(t, x).$$

If the function $f(t, x, y)$ fulfills the Lipschitz condition $(\mathcal{L})$ with respect to y in the set

$$\tilde{\mathcal{K}} = \{(t, x, y) : (t, x) \in \overline{D}, \alpha \le y \le \beta\},$$

then for all $(t, x) \in \overline{D}$ and ξ, η with $\alpha \le \xi < \eta \le \beta$ we have

$$\overline{f}(t, x, \eta) - \overline{f}(t, x, \xi) \ge -L(\eta - \xi).$$

In other words, for every $(t, x) \in \overline{D}$ the function

$$\overline{f}_L(t, x, y) := \overline{f}(t, x, y) + Ly$$

is increasing with respect to y in the interval $\langle \alpha, \beta \rangle$.

The initial-boundary value problem (4.44) is obviously equivalent to the initial-boundary value problem

$$\begin{cases} (\mathcal{D}_t - \mathcal{L}_L)[z](t, x) = \overline{\mathbf{F}}_L[z](t, x) := \overline{f}_L(t, x, z(t, x)) & \text{for } (t, x) \in D, \\ z(t, x) = 0 \quad \text{for } (t, x) \in \Gamma, \end{cases} \tag{4.45}$$

where

$$\mathcal{L}_L := \mathcal{L} - L\mathcal{I}.$$

Moreover, u_0 (v_0) is a lower (upper) solution of problem (4.44) if and only if it is a lower (upper) solution of problem (4.45); $\overline{\mathbf{F}}$ and $\overline{\mathbf{F}}_L$ have the same continuity properties and $\mathcal{L}_L$ is a differential operator of the same type as $\mathcal{L}$.

4.4 *Certain Variants of Monotone Iterative Methods*

Above, the four monotone iterative methods have been used to examine the existence of a solution for problem (4.1). Here we shall apply two other monotone iterative methods. These methods make it again possible to build sequences of successive approximations that converge to a solution sought for at a rate higher than in the case of the iterative sequences of successive approximations defined by (4.2) and (4.3). In general, it consists of the following: if we consider some nonlinear system of equations whose right-hand sides are functions of the form $f = f(t, x, s, s)$, then the successive approximation sequence $\{\tilde{u}_n\}$ arising from the iteration $f(t, x, \tilde{u}_n, \tilde{u}_{n-1})$ is considered. Thus it is a pseudo-linearization of the nonlinear problem. Applying the above iterative method to the problem here considered under appropriate assumptions on the functions f, the sequence $\{\tilde{u}_n\}$ tends to the searched for exact solution at a rate not lower than that of the successive approximation sequence $\{u_n\}$ given by the iteration $\mathcal{F}^i[u_n^i](t, x) = f^i(t, x, u_{n-1}, u_{n-1})$.

We will define the successive terms of approximation sequences $\{\tilde{u}_n\}$ and $\{\tilde{v}_n\}$ as regular solutions of the following infinite systems of equations:

$$\mathcal{F}^i\left[\tilde{u}_n^i\right](t, x) = f^i(t, x, \tilde{u}_n(t, x), \tilde{u}_{n-1}) \quad \text{for } (t, x) \in D \quad \text{and} \quad i \in S \tag{4.46}$$

and

$$\mathcal{F}\left[\tilde{v}_n^i\right](t, x) = f^i(t, x, \tilde{v}_n(t, x), \tilde{v}_{n-1}) \quad \text{for } (t, x) \in D \quad \text{and} \quad i \in S \tag{4.47}$$

for $n = 1, 2, \ldots$, satisfying the homogeneous initial boundary condition (2.6).

Theorem 4.6 *Under assumptions* A_0, (H_a), (H_f), *and conditions* (W), (L), (V) *which hold in the set* $\mathcal{K}$, *if we start the approximation process from the same ordered pair of a lower* u_0 *and an upper* v_0 *solution of problem* (4.1) *in* $\overline{D}$, *as the pair of initial iterations then the sequences* $\{\tilde{u}_n\}$ *and* $\{\tilde{v}_n\}$ *given by* (4.46), (4.47), *and* (2.6) *are well defined in* $C_S^{2+\alpha}(\overline{D})$, *the functions* $\tilde{u}_n$ *and* $\tilde{v}_n$ $(n = 1, 2, \ldots)$ *are a lower and an upper solution of problem* (4.1) *in* $\overline{D}$, *respectively, and these sequences converge monotonously and uniformly to a solution* z *of problem* (4.1) *in* $\overline{D}$ *at a rate not lower than that of the iterative sequences* $\{u_n\}$ *and* $\{v_n\}$ *defined by* (4.2), (4.3), *and* (2.6) *in* $\overline{D}$, *namely the inequalities*

$$u_0(t, x) \le \tilde{u}_n(t, x) \le \tilde{u}_{n+1}(t, x) \le \tilde{v}_{n+1}(t, x) \le \tilde{v}_n(t, x) \le v_0(t, x) \tag{4.48}$$

and

$$u_n(t, x) \le \tilde{u}_n(t, x) \le \tilde{v}_n(t, x) \le v_n(t, x) \tag{4.49}$$

hold for $(t, x) \in \overline{D}$ *and* $n = 1, 2, \ldots$ *The function*

$$z = z(t, x) = \lim_{n \to \infty} \tilde{u}_n(t, x) \quad \text{uniformly in } \overline{D}$$

is the unique global regular solution of problem (4.1) *within the sector* $\langle u_0, v_0 \rangle$, *and* $z \in C_S^{2+\alpha}(\overline{D})$.

Proof.   From Theorem 4.2 it follows that the successive terms of approximation sequences $\tilde{u}_n$ and $\tilde{v}_n$ are well defined and $\tilde{u}_n$, $\tilde{v}_n \in C_S^{2+\alpha}(\overline{D})$ for $n = 1, 2, \ldots$

To prove the theorem, it is enough to show inequality (4.49), because the other statements of the theorem are obvious. To this end we use the mathematical induction. Indeed, for $n = 1$, by (4.48), (4.46), (4.2), and (**W**) we obtain

$$\mathcal{F}^i[\tilde{u}_1^i](t, x) - \mathcal{F}^i[u_1^i](t, x)$$
$$= f^i(t, x, \tilde{u}_1(t, x), u_0) - f^i(t, x, u_0(t, x), u_0) \geq 0, \quad i \in S, \quad \text{in } D$$

with homogeneous initial-boundary condition (2.6). Hence, by virtue of Lemma 3.1, we obtain

$$\tilde{u}_1(t, x) \geq u_1(t, x) \quad \text{for } (t, x) \in \overline{D} \quad \text{and} \quad i \in S.$$

If now

$$\tilde{u}_{n-1}(t, x) \geq u_{n-1}(t, x) \quad \text{for } (t, x) \in \overline{D} \quad \text{and} \quad i \in S,$$

then by (4.46), (4.48), (4.1), and (**W**) we come to the inequalities

$$\mathcal{F}^i[\tilde{u}_n^i](t, x) - \mathcal{F}^i[u_n^i](t, x) = f^i(t, x, \tilde{u}_n(t, x), \tilde{u}_{n-1}) - f^i(t, x, u_{n-1}(t, x), u_{n-1})$$
$$\geq f^i(t, x, \tilde{u}_{n-1}(t, x), \tilde{u}_{n-1}) - f^i(t, x, u_{n-1}(t, x), u_{n-1}) \geq 0$$
$$\text{for } (t, x) \in \overline{D} \quad \text{and} \quad i \in S$$

with condition (2.6). Hence by Lemma 3.1 we obtain

$$\tilde{u}_n(t, x) \geq u_n(t, x) \quad \text{for } (t, x) \in \overline{D} \quad \text{and} \quad i \in S.$$

Analogously

$$v_n(t, x) \geq \tilde{v}_n(t, x) \quad \text{for } (t, x) \in \overline{D} \quad \text{and} \quad i \in S.$$

Therefore, by induction, inequality (4.49) is proved. This completes the proof.   □

4.5 Another Variant of the Monotone Iterative Method

Let us consider the infinite countable system of equations of the form

$$\mathcal{F}^j[z^j](t, x) = \tilde{f}^j(t, x, z) \quad \text{for } (t, x) \in D \quad \text{and} \quad j \in \mathbb{N} \tag{4.50}$$

with homogeneous initial-boundary condition (2.6).

We will define the successive terms of the approximation sequences of problem (4.50) and (2.6) as solutions of the following infinite countable systems of equations with Volterra functionals[2]

$$\mathcal{F}^j[\overline{u}_n^i](t, x) = \tilde{f}^j\left(t, x, [\overline{u}_n, \overline{u}_{n-1}]^i\right) \quad \text{for } (t, x) \in \overline{D}, \quad j \in \mathbb{N} \tag{4.51}$$

---

[2] For any $\eta, y \in \mathcal{B}(S)$ and for every fixed $i \in S$, let $[\eta, y]^i$ denote an element of $\mathcal{B}(S)$ with the description

and

$$\mathcal{F}^j\big[\bar{v}_n^j\big](t, x) = \tilde{f}^j\left(t, x, [\bar{v}_n, \bar{v}_{n-1}]^j\right) \quad \text{for } (t, x) \in \overline{D}, \quad j \in \mathbb{N} \tag{4.52}$$

for $n = 1, 2, \ldots$ in D, with initial-boundary condition (2.6).

Theorem 4.7 *Let assumptions* $\mathbf{A_0}$, $(\mathbf{H_a})$, $(\mathbf{H_f})$ *hold, and conditions* (**W**), (**L**), (**V**) *hold in the set* $\tilde{\mathcal{K}}$. *If we define the successive terms of approximation sequences* $\{\bar{u}_n\}$ *and* $\{\bar{v}_n\}$ *as regular solutions in* $\overline{D}$ *of systems* (4.51) *and* (4.52) *with homogeneous initial-boundary condition* (2.6) *and if* $\bar{u}_0 = u_0, \bar{v}_0 = v_0$, *then*

i. $\{\bar{u}_n\}, \{\bar{v}_n\}$ *are well defined and* $\bar{u}_n, \bar{v}_n \in C_{\mathbb{N}}^{2+\alpha}(\overline{D})$ *for* $n = 1, 2, \ldots$;
ii. *the inequalities*

$$u_0(t, x) \le \bar{u}_n(t, x) \le \bar{u}_{n+1}(t, x) \le \bar{v}_{n+1}(t, x) \le \bar{v}_n(t, x) \le v_0(t, x) \tag{4.53}$$

for $(t, x) \in \overline{D}$ *and* $n = 1, 2, \ldots$ *hold;*
iii. *the functions* $\bar{u}_n$ *and* $\bar{v}_n$ *for* $n = 1, 2, \ldots$ *are lower and upper solutions of problem* (4.1) *in* $\overline{D}$, *respectively;*
iv. *if we start two approximation processes from the same pair of a lower solution* u_0 *and an upper solution* v_0 *of problem* (4.50) *and* (2.6) *in* $\overline{D}$, *then the sequences* $\{\bar{u}_n\}$ *and* $\{\bar{v}_n\}$ *converge monotonously and uniformly to a solution* z *of problem* (4.50) *and* (2.6) *in* $\overline{D}$ *at a rate not lower than that of the iterative sequences* $\{u_n\}$ *and* $\{v_n\}$ *defined by* (4.2), (4.3), *and* (2.6) *in* $\overline{D}$, *i.e., the inequalities*

$$u_n(t, x) \le \bar{u}_n(t, x) \le \bar{v}_n(t, x) \le v_n(t, x) \tag{4.54}$$

for $(t, x) \in \overline{D}$ *and* $n = 1, 2, \ldots$ *hold;*
v. *the function* $z = z(t, x) = \lim_{n \to \infty} \bar{u}_n(t, x)$ *uniformly in* $\overline{D}$ *is the unique global regular solution of problem* (4.50) *and* (2.6) *within the sector* $\langle u_0, v_0 \rangle$, *and* $z \in C_{\mathbb{N}}^{2+\alpha}(\overline{D})$.

Proof. Using the auxiliary lemmas we prove the theorem by induction. The proof is simple and similar for the lower and upper solutions, so in both cases we present the proof for lower solutions only.

Since $u_0 \in C_{\mathbb{N}}^{0+\alpha}(\overline{D})$, then from Lemma 4.1 and the theorem on the existence and uniqueness of the solution of the Fourier first initial-boundary value problem for the finite system of equations, it follows that there exists the regular unique solution $\bar{u}_1$ of problem (4.51) and

$$[\eta, y]^i := \begin{cases} y^j & \text{for all } j \ne i, \quad j \in \mathbb{N}, \\ \eta^i & \text{for } j = i. \end{cases}$$

In the case if $S = \mathbb{N}$ then $\eta, y \in l^\infty$ and

$$[\eta, y]^i := (y^1, y^2, \ldots, y^{i-1}, \eta^i, y^{i+1}, \ldots).$$

(2.6) in $\overline{D}$ and $\overline{u}_1 \in C_{\mathbb{N}}^{2+\alpha}(\overline{D})$. Analogously, if $\overline{u}_{n-1} \in C_{\mathbb{N}}^{2+\alpha}(\overline{D})$, then there exists the regular unique solution $\overline{u}_n$ of problem (4.52) and (2.6) in $\overline{D}$, $\overline{u}_n \in C_{\mathbb{N}}^{2+\alpha}(\overline{D})$ and (i) is proved by induction.

Since u_0 is a lower solution, it satisfies the inequalities

$$\mathcal{F}^j[u_0^j](t, x) \le \tilde{f}^j(t, x, u_0) = \tilde{f}^j\left(t, x, [u_0, u_0]^j\right) \quad \text{for } (t, x) \in D \text{ and } j \in \mathbb{N}$$

with condition (2.6). The function $\overline{u}_1$ is a solution of the equation

$$\mathcal{F}^j[\overline{u}_1^j](t, x) = \tilde{f}^j\left(t, x, [\overline{u}_1, u_0]^j\right) \quad \text{for } (t, x) \in D \text{ and } j \in \mathbb{N}$$

with condition (2.6). Therefore, by Lemma 3.1 we obtain

$$u_0(t, x) \le \overline{u}_1(t, x) \quad \text{for } (t, x) \in \overline{D}. \tag{4.55}$$

Moreover, by (4.55) and (**W**) there is

$$\mathcal{F}^j[\overline{u}_1^j](t, x) = \tilde{f}^j\left(t, x, [\overline{u}_1, u_0]^j\right) \le \tilde{f}^j\left(t, x, [\overline{u}_1, \overline{u}_1]^j\right)$$
$$\text{for } (t, x) \in D, \quad \text{and} \quad j \in \mathbb{N},$$

so the function $\overline{u}_1$ is a lower solution of problem (4.50) and (2.6) in $\overline{D}$.

Analogously, if $\overline{u}_{n-1}$ is a lower solution, then by (2.20) we obtain

$$\mathcal{F}^j[\overline{u}_{n-1}^j](t, x) \le \tilde{f}^j(t, x, \overline{u}_{n-1}) \quad \text{for } (t, x) \in D \text{ and } j \in \mathbb{N}$$

and $\overline{u}_n$ is a solution of system (4.51) with condition (2.6). Therefore, by Lemma 3.1 we obtain

$$\overline{u}_{n-1}(t, x) \le \overline{u}_n(t, x) \quad \text{for } (t, x) \in \overline{D}. \tag{4.56}$$

Moreover, by (4.56) and (**W**) there is

$$\mathcal{F}^j[\overline{u}_n^j](t, x) = \tilde{f}^j\left(t, x, [\overline{u}_n, \overline{u}_{n-1}]^j\right) \le \tilde{f}^j\left(t, x, [\overline{u}_n, \overline{u}_n]^j\right) = \tilde{f}^j(t, x, \overline{u}_n),$$
$$\text{for } (t, x) \in D \text{ and } j \in \mathbb{N},$$

so the function $\overline{u}_n$ is a lower solution of problem (4.50) and (2.6) in $\overline{D}$.

From inequalities (4.55) and (4.56) for lower solutions and the analogous inequalities for upper solutions, using mathematical induction, we obtain inequality (4.53).

We now prove inequality (4.54) by induction. From (4.2) and (4.56) for $n = 1$, (4.55) and (**W**) we obtain

$$\mathcal{F}^j[\overline{u}_1^j](t, x) - \mathcal{F}^j[u_1^j](t, x) = \tilde{f}^j\left(t, x, [\overline{u}_1, u_0]^j\right) - \tilde{f}^j\left(t, x, [u_0, u_0]^j\right) \ge 0$$
$$\text{for } (t, x) \in D \text{ and } j \in \mathbb{N}$$

with condition (2.6). By virtue of Lemma 3.1 we obtain

$$\bar{u}_1(t, x) \geq u_1(t, x) \quad \text{for } (t, x) \in \overline{D}. \tag{4.57}$$

Let now

$$\bar{u}_{n-1}(t, x) \geq u_{n-1}(t, x) \quad \text{for } (t, x) \in \overline{D}, \tag{4.58}$$

then by $(\mathscr{W})$, because u_{n-1} is a lower solution of problem (4.50) and (2.6) in $\overline{D}$, we derive

$$\mathcal{F}^j\left[u_{n-1}^i\right](t, x) \leq \tilde{f}^j\left(t, x, [u_{n-1}, u_{n-1}]^i\right) \leq \tilde{f}^j\left(t, x, [u_{n-1}, \bar{u}_{n-1}]^j\right)$$
$$\text{for } (t, x) \in D \quad \text{and} \quad j \in \mathbb{N}$$

with condition (2.6). Therefore, by (4.51) and Lemma 3.1 we have

$$\bar{u}_n(t, x) \geq u_{n-1}(t, x) \quad \text{for } (t, x) \in \overline{D}. \tag{4.59}$$

From (4.2), (4.51), (4.58), and (4.59) we get

$$\mathcal{F}^j\left[\bar{u}_n^j\right](t, x) - \mathcal{F}^j\left[u_n^j\right](t, x) = \tilde{f}^j\left(t, x, [\bar{u}_n, \bar{u}_{n-1}]^j\right) - \tilde{f}^j\left(t, x, [u_n, u_{n-1}]^j\right)$$
$$\geq \tilde{f}^j\left(t, x, [\bar{u}_n, u_{n-1}]^j\right) - \tilde{f}^j\left(t, x, [u_n, u_{n-1}]^j\right) \geq 0$$
$$\text{for } (t, x) \in \overline{D} \quad \text{and} \quad j \in \mathbb{N}$$

with condition (2.6). By virtue of Lemma 3.1 we obtain

$$\bar{u}_n(t, x) \geq u_n(t, x) \quad \text{in } \overline{D}. \tag{4.60}$$

From (4.53), (4.57), and (4.60) we finally obtain inequality (4.54).

By (4.54) and (4.15) there is

$$\bar{v}_n^j(t, x) - \bar{u}_n^j(t, x) \leq N_0 \frac{(Lt)^n}{n!}, \quad j \in \mathbb{N}, \quad n = 1, 2, \ldots, \tag{4.61}$$

where $N_0 := \|v_0 - u_0\|_0 = \text{const} < \infty$ for $(t, x) \in \overline{D}$.

As a direct consequence, we obtain

$$\lim_{n \to \infty}\left[\bar{v}_n^j(t, x) - \bar{u}_n^j(t, x)\right] = 0 \quad \text{uniformly in } \overline{D}, \quad j \in \mathbb{N}.$$

By arguments similar to that used in Section 4.1 we show that

$$z = z(t, x) := \lim_{n \to \infty} \bar{u}_n(t, x) \quad \text{uniformly in } \overline{D} \tag{4.62}$$

is the unique global regular solution of problem (4.50) and (2.6) within the sector $\langle u_0, v_0\rangle$. Obviously, $z \in C_{\mathbb{N}}^{2+\alpha}(\overline{D})$. $\qquad\square$

Remark 4.4 It is easy to see that the method used in the proof does not need the assumption that the considered system is countable. Therefore, in Theorem 4.7 a countable infinite system may be replaced by an arbitrary infinite system.

4.6 Method of Direct Iterations in Unbounded Domains

We consider an infinite system of equations of the form (2.2) with initial-boundary condition (2.5), i.e., the problem

$$\mathcal{F}^i[z^i](t, x) = \tilde{f}^i(t, x, z) \quad \text{for } (t, x) \in \Omega \quad \text{and} \quad i \in S, \tag{4.63}$$

$$z(t, x) = \phi(t, x) \quad \text{for } (t, x) \in \Gamma_\Omega, \tag{4.64}$$

where Ω is an arbitrary open domain in the time-space $\mathbb{R}^{m+1}$, unbounded with respect to x, and the operators $\mathcal{F}^i, i \in S$, are uniformly parabolic in $\overline{\Omega}$, where Γ_Ω is the parabolic boundary of Ω and $\overline{\Omega} := \Omega \cup \Gamma_\Omega$.

Now we assume that the open and unbounded (with respect to x) domain, $\Omega \subset \mathbb{R}^{m-1}$, has property $\mathcal{P}$. The boundary $\partial\Omega$ of domain Ω consists of m-dimensional (bounded or unbounded) domains S_0 (where S_0 is a nonempty set) and S_T lying in the hyperplanes $t = 0$ and $t = T$, respectively, and a certain surface σ_Ω (not necessarily connected) of a class $C^{2+\alpha}$, i.e., it consists of a finite number of surfaces of a class $C^{2+\alpha}$, not overlapping but having common boundary points lying in the zone $0 < t < T$, which is not tangent to any hyperplane $t = \text{const}$. We denote $\Gamma_\Omega := S_0 \cup \sigma_\Omega$. We assume that the number T is sufficiently small. Precisely, $T \leq h_0 = \text{const}$, where h_0 is a positive constant depending on problem (4.64).

We denote by Ω_R the part of the domain Ω contained inside the cylindric surface Σ_R described by the equation $\sum_{j=1}^n x_j^2 = R^2$ and $\Gamma_{\Omega_R} := \partial\Omega_R \backslash S_T$.

Moreover, for an arbitrary fixed $\tau, 0 < \tau \leq T$, we define:

$$\Omega^\tau := \Omega \cap \{(t, x) : 0 < t \leq \tau, x \in \mathbb{R}^m\},$$

$$\sigma_\Omega^\tau := \sigma_\Omega \cap \{(t, x) : 0 < t \leq \tau, x \in \mathbb{R}^m\},$$

$$\Gamma^\tau := S_0 \cup \sigma_\Omega^\tau, \quad \overline{\Omega}^\tau := \Omega^\tau \cup \Gamma^\tau.$$

Obviously, $\Omega^T = \Omega$.

Analogously as in Section 2.2, we define the Banach space $\mathcal{B}(S)$ and denote by $C_S(\overline{\Omega})$ the Banach space of mappings

$$w : \overline{\Omega} \to \mathcal{B}(S), \quad (t, x) \mapsto w(t, x)$$

and

$$w(t, x) : S \to \mathbb{R}, \quad i \mapsto w^i(t, x),$$

where the functions w^i are continuous in $\overline{\Omega}$, with the finite norm

$$\|w\|_0 := \sup \left\{ \left| w^i(t, x) \right| : (t, x) \in \overline{\Omega}, i \in S \right\}.$$

In the Banach space $C_S(\overline{\Omega})$ the partial order is defined by means of the positive cone

$$C_S^+(\overline{\Omega}) := \{w : w = \{w^i\}_{i \in S} \in C_S(\overline{\Omega}), w^i(t, x) \geq 0 \text{ for } (t, x) \in \overline{\Omega}, i \in S\}.$$

In the theory of parabolic equations it is well known that initial-value problems in unbounded spatial domains are considered with the growth condition $|w(t, x)| \leq M \exp(K |x|^2)$; otherwise, these problems are ill posed.

Therefore, we introduce the *class E_2 of real functions* $w = w(t, x)$ for which there exist positive constants M and K such that the following growth condition is fulfilled

$$|w(t, x)| \leq M \exp(K |x|^2) \quad \text{for } (t, x) \in \Omega.$$

We then say that $w \in E_2(M, K; \Omega)$ or shortly $w \in E_2$.

Denote by $C_{S, E_2}(\overline{\Omega})$ a Banach space of mappings $w \in C_S(\overline{\Omega})$ belonging to the class $E_2(M, K; \Omega)$ with the finite *weighted norm*

$$\|w\|_0^{E_2} := \sup \left\{ \left|w^i(t, x)\right| \exp(-K|x|^2) : (t, x) \in \overline{\Omega}, \ i \in S \right\}.$$

For $w \in C_{S, E_2}(\overline{\Omega})$ and for a fixed $t, 0 \leq t \leq T$, we define

$$\|w\|_{0,t}^{E_2} := \sup \left\{ \left|w^i(\tilde{t}, \tilde{x})\right| \exp(-K|\tilde{x}|^2) : (\tilde{t}, \tilde{x}) \in \overline{\Omega}^t, \ i \in S \right\}.$$

A mapping $w \in C_{S, E_2}(\overline{\Omega})$ will be called regular in $\overline{\Omega}$ if the functions $w^i, i \in S$, have continuous derivatives $\mathcal{D}_t w^i, \mathcal{D}_{x_j} w^i, \mathcal{D}_{x_j x_k}^2 w^i$ in Ω for $j, k = 1, 2, \ldots, m$ (i.e., $w \in C_{S, E_2}^{\text{reg}}(\overline{\Omega})$).

We define the Hölder space $C^{k+\alpha}(\overline{\Omega})$ and by $C_{S, E_2}^{k+\alpha}(\overline{\Omega})$ we denote a Banach space of mappings $w = \{w^i\}_{i \in S}$ with the finite weighted norm

$$\|w\|_{k+\alpha}^{E_2} := \sup \left\{ |w^i|_{k+\alpha} \exp(-K |x|^2) : (t, x) \in \overline{\Omega}, \ i \in S \right\},$$

where $K > 0$ is a constant.

Assumption $\tilde{H}_a$. We assume that the coefficients $a_{jk}^i = a_{jk}^i(t, x)$, $a_{jk}^i = a_{kj}^i$, $b_j^i = b_j^i(t, x)$ $(j, k = 1, 2, \ldots, m, i \in S)$ of the operators $\mathcal{L}^i, i \in S$, are continuous with respect to t and x in $\overline{\Omega}$, bounded and locally Hölder continuous with exponent α $(0 < \alpha < 1)$ with respect to t and x in $\overline{\Omega}$ and their Hölder norms are uniformly bounded.

We assume that the functions

$$\tilde{f}^i : \overline{\Omega} \times C_S(\overline{\Omega}) \to \mathbb{R} \quad (t, x, s) \mapsto \tilde{f}^i(t, x, s), \quad i \in S$$

are continuous and satisfy the following assumptions:

Assumption $\tilde{H}_f$. The functions $\tilde{f}^i$, $i \in S$, are *locally Hölder continuous* with exponent α $(0 < \alpha < 1)$ with respect to t and x in $\overline{\Omega}$, and their Hölder norms are uniformly bounded.

Assumption E_f. $\tilde{f}^i(t, x, 0) \in E_2(M_f, K_f; \Omega)$, $i \in S$.

Condition $\tilde{L}_{E_2}$. The functions $\tilde{f}^i(t, x, s)$, $i \in S$, fulfill the *Lipschitz condition* $(\tilde{L}_{E_2})$ with respect to s, i.e., for arbitrary $s, \tilde{s} \in C_{S, E_2}(\overline{\Omega})$ there is

$$\left| \tilde{f}^i(t, x, s) - \tilde{f}^i(t, x, \tilde{s}) \right| \leq L_1 \|s - \tilde{s}\|_0^{E_2} \quad \text{for } (t, x) \in \Omega,$$

where $L_1 > 0$ is a constant.

Condition $\tilde{L}_{*}_{E_2}$. We say that the functions $\tilde{f}^i(t, x, s)$, $i \in S$, fulfill the *Lipschitz-Volterra condition* $(\tilde{L}_{*}_{E_2})$ with respect to s if for arbitrary $s, \tilde{s} \in C_{S, E_2}(\overline{\Omega})$ there is

$$\left| \tilde{f}^i(t, x, s) - \tilde{f}^i(t, x, \tilde{s}) \right| \leq L_2 \|s - \tilde{s}\|_{0, t}^{E_2} \quad \text{for } (t, x) \in \Omega,$$

where $L_2 > 0$ is a constant.

Moreover, we assume that

Assumption $\tilde{H}_\phi$. $\phi \in C_S^{2+\alpha}(\Gamma_\Omega)$, where $0 < \alpha < 1$.

Assumption E_ϕ. $\phi \in E_2(M_\phi, K_\phi; \Gamma_\Omega)$.

Functions $u, v \in C_{S, E_2}^{\text{reg}}(\overline{\Omega})$ satisfying the infinite systems of inequalities

$$\begin{cases} \mathcal{F}^i[u^i](t, x) \leq \tilde{f}^i(t, x, u) & \text{for } (t, x) \in \Omega, \\ u^i(t, x) \leq \phi^i(t, x) & \text{for } (t, x) \in \Gamma_\Omega \text{ and } i \in S, \end{cases} \tag{4.65}$$

$$\begin{cases} \mathcal{F}^i[v^i](t, x) \geq \tilde{f}^i(t, x, v) & \text{for } (t, x) \in \Omega, \\ v^i(t, x) \geq \phi^i(t, x) & \text{for } (t, x) \in \Gamma_\Omega \text{ and } i \in S \end{cases} \tag{4.66}$$

are called, respectively, a lower and an upper solution of problem (4.63), (4.64) in unbounded domain $\overline{\Omega}$.

Assumption $\tilde{A}$. We assume that there exists at least one pair u_0 and v_0 of a lower and an upper solution of problem (4.63), (4.64) in $\overline{\Omega}$ and $u_0, v_0 \in C_S^{0+\alpha}(\overline{\Omega})$.

A pair of a lower u_0 and an upper solution v_0 of problem (4.63), (4.64) in $\overline{D}$ is called an *ordered pair* if $u_0 \leq v_0$ in $\overline{D}$.

The inequality $u_0 \leq v_0$ does not follow directly from inequalities (4.65) and (4.66). Therefore, we adopt the following fundamental assumption.

Assumption $\tilde{A}_0$. We assume that there exists at least one ordered pair u_0 and v_0 of a lower and an upper solution of problem (4.63), (4.64) in $\overline{\Omega}$, and $u_0, v_0 \in C_S^{0+\alpha}(\overline{\Omega})$.

Let $\beta \in C_S(\overline{\Omega})$ be a sufficiently regular function. Denote by $\tilde{\mathcal{P}}$ the operator

$$\tilde{\mathcal{P}}: \beta \mapsto \tilde{\mathcal{P}}[\beta] = \gamma,$$

where γ is the (supposedly unique) solution of the initial-boundary value problem

$$\begin{cases} \mathcal{F}^i[\gamma^i](t, x) = \tilde{f}^i(t, x, \beta) & \text{for } (t, x) \in \Omega, \\ \gamma^i(t, x) = \phi^i(t, x) & \text{for } (t, x) \in \Gamma_\Omega \text{ and } i \in S. \end{cases} \tag{4.67}$$

The operator $\tilde{\mathcal{P}}$ is the composition of the nonlinear operator $\tilde{\mathbf{F}} = \{\tilde{\mathbf{F}}^i\}_{i \in S}$ generated by the functions $\tilde{f}^i(t, x, s)$, $i \in S$, and defined for any $\beta \in C_S(\overline{\Omega})$ as follows.

$$\tilde{\mathbf{F}}^i : \beta \mapsto \tilde{\mathbf{F}}^i[\beta] = \delta^i,$$

where

$$\tilde{\mathbf{F}}^i[\beta](t, x) := \tilde{f}^i(t, x, \beta) = \delta^i(t, x) \quad \text{for } (t, x) \in \Omega \text{ and } i \in S, \tag{4.68}$$

and the operator

$$\mathcal{G}: \delta \mapsto \mathcal{G}[\delta] = \gamma,$$

where γ is the (supposedly unique) solution of the linear problem

$$\begin{cases} \mathcal{F}^i[\gamma^i](t, x) = \delta^i(t, x) & \text{for } (t, x) \in \Omega, \\ \gamma^i(t, x) = \phi^i(t, x) & \text{for } (t, x) \in \Gamma_\Omega \text{ and } i \in S. \end{cases} \tag{4.69}$$

Hence

$$\tilde{\mathcal{P}} = \mathcal{G} \circ \tilde{\mathbf{F}}.$$

Lemma 4.8 *If $\beta \in C^{0+\alpha}_{S, E_2}(\overline{\Omega})$ and the function $\tilde{f} = \{\tilde{f}^i\}_{i \in S}$ generating the nonlinear operator satisfies conditions $(\tilde{H}_f)$, (E_f), and $(\tilde{L}_{*_{E_2}})$, then*

$$\tilde{\mathbf{F}}: C^{0+\alpha}_{S, E_2}(\overline{\Omega}) \to C^{0+\alpha}_{S, E_2}(\overline{\Omega}), \quad \beta \mapsto \tilde{\mathbf{F}}[\beta] = \delta.$$

Proof. Using the same argument as in the proof of Lemma 4.3 we prove that $\delta \in C^{0+\alpha}_S(\overline{\Omega})$ and by (E_f) and $(\tilde{\mathcal{L}}_{E_2})$ there is $\delta(t, x) \in E_2(M_\delta, K_\delta; \Omega)$. This completes the proof. $\qquad\square$

Lemma 4.9 *If assumptions $(\tilde{H}_a)$, $(\tilde{H}_f)$, (E_f), $(\tilde{H}_\phi)$, (E_ϕ), and $(\tilde{L}_{E_2})$ hold, then the operator $\tilde{\mathcal{P}}$ is well defined for $\beta \in C^{0+\alpha}_{S, E_2}(\overline{\Omega})$, where $T \le h_0$ and*

$$\tilde{\mathcal{P}}: C^{0+\alpha}_{S, E_2}(\overline{\Omega}) \to C^{2+\alpha}_{S, E_2}(\overline{\Omega}), \quad \beta \mapsto \tilde{\mathcal{P}}[\beta] = \gamma.$$

Proof. Observe that system (4.67) has the following property: the ith equation depends on the ith unknown function only. Therefore, its solution is a collection of separate solutions of these scalar problems for linear equations considered in a suitable space. This fact, Lemma 4.8, and assumptions on the domain Ω imply that the theorem on the existence and uniqueness of a solution for a linear parabolic problem in an unbounded domain holds. Therefore, problem (4.63), (4.64) has exactly one regular solution, $\gamma \in C_{S,E_2}^{2+\alpha}(\overline{\Omega})$, provided that $T \leq h_0$. □

Using the same arguments as in Section 4.1, we may prove the following lemma.

Lemma 4.10 *Let all the assumptions of Lemmas 4.8 and 4.9 hold, β be an upper solution and α be a lower solution of problem (4.63), (4.64) in $\overline{\Omega}$, α, $\beta \in \langle u_0, v_0 \rangle$ and condition* (**W**) *hold. Then*

$$\alpha(t, x) \leq \tilde{P}[\beta](t, x) \leq \beta(t, x) \quad in \ \overline{\Omega} \tag{4.70}$$

and $\gamma = \tilde{P}[\beta]$ is an upper solution of problem (4.63) and (4.64) in $\overline{\Omega}$. Analogously

$$\alpha(t, x) \leq \tilde{P}[\alpha](t, x) \leq \beta(t, x) \quad in \ \overline{\Omega} \tag{4.71}$$

and $\eta = \tilde{P}[\alpha]$ is a lower solution of problem (4.63), (4.64) in $\overline{\Omega}$.

Lemma 4.11 *If the assumptions of Lemma 4.9 and condition* (**W**) *hold, then the operator $\tilde{P}$ is monotone increasing (isotone).*

Proof.   Of course, $\tilde{P}$ is monotone increasing, because $\tilde{F}$ is monotone increasing by assumption (**W**) and $\mathcal{G}$ is monotone increasing operator by the maximum principle. This completes the proof. □

Theorem 4.8 *Let assumptions $\tilde{A}_0$, $(\tilde{H}_a)$, $(\tilde{H}_f)$, (E_f), $(\tilde{H}_\phi)$, (E_ϕ) hold, and conditions $(\tilde{L}_{E_2})$, (**W**), $(\tilde{L}_{\underset{E_2}{*}})$ hold in the set*

$$\tilde{\mathcal{K}} := \{(t, x, s) : (t, x) \in \overline{\Omega}, \quad s \in \langle u_0, v_0 \rangle \},$$

where

$$\langle u_0, v_0 \rangle := \left\{ w \in C_{S,E_2}(\overline{\Omega}) : u_0(t, x) \leq w(t, x) \leq v_0(t, x) \ for \ (t, x) \in \overline{\Omega} \right\}.$$

Consider the following infinite systems of linear equations:

$$\mathcal{F}^i[u_n^i](t, x) = \tilde{f}^i(t, x, u_{n-1}) \quad for \ (t, x) \in \Omega, \quad and \ i \in S \tag{4.72}$$

and

$$\mathcal{F}^i[v_n^i](t, x) = \tilde{f}^i(t, x, v_{n-1}) \quad for \ (t, x) \in \Omega, \quad and \ i \in S \tag{4.73}$$

for $n = 1, 2, \ldots$ with initial-boundary condition

$$u_n(t, x) = \phi(t, x) \quad for \ (t, x) \in \Gamma_\Omega$$

and

$$v_n(t, x) = \phi(t, x) \quad for \ (t, x) \in \Gamma_\Omega,$$

respectively, and let $\tilde{N}_0 = \|v_0 - u_0\|_0^{E_2} < \infty$, where an ordered pair of a lower u_0 and an upper v_0 solution will be the pair of initial iteration in this iterative process.

Then

i. *there exist unique solutions u_n and v_n for $n = 1, 2, \ldots$ of systems (4.72) and (4.73) with suitable initial-boundary conditions in $\overline{\Omega}$ and u_n, $v_n \in C^{2+\alpha}_{S, E_2}(\overline{\Omega})$;*

ii. *the inequalities*

$$u_0(t, x) \le u_n(t, x) \le u_{n+1}(t, x) \le v_{n+1}(t, x) \le v_n(t, x) \le v_0(t, x) \tag{4.74}$$

for $(t, x) \in \overline{\Omega}$ and $n = 1, 2, \ldots$ hold;

iii. *the functions u_n and v_n for $n = 1, 2, \ldots$ are a lower and an upper solution of problem (4.63), (4.64) in $\overline{\Omega}$, respectively;*

iv. *the following estimate:*

$$w_n^i(t, x) \le \tilde{N}_0 \frac{(Lt)^n}{n!} \quad \text{for } (t, x) \in \overline{\Omega}, \quad \text{and} \quad n = 1, 2, \ldots, \quad i \in S, \tag{4.75}$$

holds, where

$$w_n^i(t, x) := v_n^i(t, x) - u_n^i(t, x) \ge 0 \quad \text{for } (t, x) \in \overline{\Omega}; \quad \text{and} \quad i \in S,$$

v. $\lim_{n \to \infty}\left[v_n^i(t, x) - u_n^i(t, x)\right] = 0$ *almost uniformly in Ω, $i \in S$;*

vi. *the function*

$$z = z(t, x) = \lim_{n \to \infty} u_n(t, x)$$

is the unique regular solution of problem (4.63), (4.64) within the sector $\langle u_0, v_0 \rangle$, and $z \in C^{2+\alpha}_{S, E_2}(\overline{\Omega})$.

Proof. Starting from the lower solution u_0 and the upper solution v_0 of problem (4.63), (4.64), we define by induction the successive terms of the iteration sequences $\{u_n\}$, $\{v_n\}$ as solutions of systems (4.72) and (4.73) of linear equations with initial-boundary condition (4.64) in $\overline{\Omega}$, or shortly

$$u_n = \tilde{\mathcal{P}}[u_{n-1}], \quad v_n = \tilde{\mathcal{P}}[v_{n-1}] \quad \text{for } n = 1, 2, \ldots.$$

From Lemmas 4.8–4.10 it follows that u_n and v_n, for $n = 1, 2, \ldots$ exist, u_n, $v_n \in C^{2+\alpha}_{S, E_2}(\overline{\Omega})$ and they are the lower and the upper solution of problem (4.63), (4.64) in $\overline{\Omega}$, respectively.

By induction, from Lemma 4.10, we derive

$$u_{n-1}(t, x) \le \tilde{\mathcal{P}}[u_{n-1}](t, x) = u_n(t, x), \tag{4.76}$$

$$v_n(t, x) = \tilde{\mathcal{P}}[v_{n-1}](t, x) \le v_{n-1}(t, x), \tag{4.77}$$

$$\text{for } (t, x) \in \Omega \quad \text{and} \quad n = 1, 2, \ldots.$$

Therefore, inequalities (4.74) hold.

From (4.74) and Assumption $\tilde{A}_0$ it follows that $u_n, v_n \in E_2(M_0, N_0; \Omega)$ for $n = 1, 2, \ldots$.

We now show by induction that (4.75) holds. It is obvious that (4.75) holds for w_0. Suppose it holds for w_n. Since the functions $\tilde{f}^i$, $i \in S$, satisfy the Lipschitz-Volterra condition $(L_{E_2}^*)$, by (4.72)–(4.75), we obtain

$$\mathcal{F}^i[w_{n+1}^i](t, x) = \tilde{f}^i(t, x, v_n) - \tilde{f}^i(t, x, u_n) \le L \, \|w_n\|_{0,t}^{E_2}.$$

By the definition of the norm $\|\cdot\|_{0,t}^{E_2}$ and by (4.75) we obtain

$$\|w_n\|_{0,t}^{E_2} \le \tilde{N}_0 \frac{(Lt)^n}{n!},$$

so we finally obtain

$$\mathcal{F}^i[w_{n+1}^i](t, x) \le \tilde{N}_0 \frac{L^{n+1} t^n}{n!} \quad \text{for } (t, x) \in \Omega, \quad \text{and} \quad i \in S \tag{4.78}$$

and

$$w_{n+1}^i(t, x) = 0 \quad \text{for } (t, x) \in \Gamma_\Omega. \tag{4.79}$$

Consider the comparison system

$$\mathcal{F}^i[M_{n+1}^i](t, x) = \tilde{N}_0 \frac{L^{n+1} t^n}{n!} \quad \text{for } (t, x) \in \Omega, \quad \text{and} \quad i \in S \tag{4.80}$$

with the initial-boundary condition

$$M_{n+1}^i(t, x) \ge 0 \quad \text{for } (t, x) \in \Gamma_\Omega. \tag{4.81}$$

It is obvious that the functions

$$M_{n+1}^i(t, x) = \tilde{N}_0 \frac{(Lt)^{n+1}}{(n+1)!} \quad \text{for } i \in S$$

are regular solutions of (4.80) and (4.81) in $\overline{\Omega}$.

Applying the theorem on differential inequalities of parabolic type in an unbounded domain to systems (4.78) and (4.80) we get

$$w_{n+1}^i(t, x) \le M_{n+1}^i(t, x) = \tilde{N}_0 \frac{(Lt)^{n+1}}{(n+1)!} \quad \text{for } (t, x) \in \Omega, \quad \text{and} \quad i \in S \tag{4.82}$$

so the induction step is proved by inequality (4.75).

As a direct consequence of (4.75) we obtain

$$\lim_{n\to\infty} \left[v_n^i(t, x) - u_n^i(t, x) \right] = 0 \quad \text{almost uniformly in } \Omega, \quad i \in S. \tag{4.83}$$

The iteration sequences $\{u_n\}$ and $\{v_n\}$ are monotone and bounded, and (4.83) holds, so there is a continuous function $U = U(t, x)$ in $\overline{\Omega}$ such that

$$\lim_{n\to\infty} u_n^i(t, x) = U^i(t, x), \quad \lim_{n\to\infty} v_n^i(t, x) = U^i(t, x) \tag{4.84}$$

almost uniformly in Ω, $i \in S$, and this function satisfies initial-boundary condition (4.64).

To prove that the function $U(t, x)$ defined by (4.84) is the regular solution of system (4.63), (4.64) in Ω, it is enough to show that it fulfills (4.63), (4.64) in any compact set contained in Ω.

Consequently, because of the definition of Ω_R, we only need to prove that it is a regular solution in Ω_R for any $R > 0$.

Since the functions $\tilde{f}^i$, $i \in S$, are monotone (condition (**W**)) and (4.74) holds, it follows that the functions $\tilde{f}^i(t, x, u_{n-1})$ $i \in S$ are uniformly bounded in Ω with respect to n.

On the basis of results concerning the properties of weak singular integrals, by means of which the solution of the linear system of equations is expressed

$$\mathcal{F}^i\left[u_n^i\right](t, x) - \tilde{f}^i(t, x, u_{n-1}) \quad \text{for } (t, x) \in \Omega_R, \quad \text{and } i \in S, \tag{4.85}$$

we conclude that the function $u_n(t, x)$ locally satisfies the Lipschitz condition with respect to x, with a constant independent of n. Hence by (4.84), we conclude that the boundary function $U(t, x)$ locally satisfies the Lipschitz condition with respect to variable x.

If we now take the system of equations

$$\mathcal{F}^i[z^i](t, x) = \tilde{f}^i(t, x, U) \quad \text{for } (t, x) \in \Omega_R, \quad \text{and } i \in S \tag{4.86}$$

with the initial-boundary condition

$$z(t, x) = U(t, x) \quad \text{for } (t, x) \in \Gamma_{\Omega_R} \tag{4.87}$$

then the last property of $U(t, x)$ together with conditions $(\mathcal{H}_f)$ and $(\mathcal{L})$ implies that the right-hand sides of system (4.86) are continuous with respect to t, x in Ω_R and locally Hölder continuous with respect to x.

Hence, in Lemma 4.9, there exists the unique regular solution z of problem (4.86) and (4.87) in $\overline{\Omega}_R$ and $z \in C_{S, E_2}^{2+\alpha}(\overline{\Omega}_R)$.

On the other hand, using (4.84), we conclude that the right-hand sides of (4.85) converge uniformly in Ω_R to the right-hand sides of (4.86),

$$\lim_{n\to\infty} f^i(t, x, u_n) = f^i(t, x, U) \quad \text{uniformly in } \overline{\Omega}_R \quad i \in S. \tag{4.88}$$

Moreover, the boundary values of $u_n(t, x)$ converge uniformly on Γ_{Ω_R} to the respective values of $U(t, x)$. Hence, using the theorem on the continuous dependence of the solution on the right-hand sides of the system and on the initial-boundary conditions to systems (4.85) and (4.86), we obtain

$$\lim_{n\to\infty} u_n^i(t, x) = z^i(t, x) \quad \text{uniformly in } \overline{\Omega}_R \quad i \in S. \tag{4.89}$$

By (4.84) and (4.89), there is

$$z = z(t, x) = U(t, x) \quad \text{for } (t, x) \in \overline{\Omega},$$

for an arbitrary R, which means that

$$z = z(t, x) = U(t, x) \quad \text{for } (t, x) \in \overline{\Omega}$$

i.e., z is the regular solution of problem (4.63), (4.64) within the sector $\langle u_0, v_0 \rangle$, and $z \in C^{2+\alpha}_{S, E_2}(\overline{\Omega})$.

The uniqueness of the solution of this problem follows directly from the uniqueness criterion of Szarski (Theorem 3.6), which ends the proof. $\qquad\square$

CHAPTER 5

Methods of Lower and Upper Solutions

5.1 Some Remarks in Connection with Applications of Numerical Methods

For certain numerical methods, it is essential that the derivatives $\mathcal{D}_t w^i$ ($i \in S$) of the functions $w = \{w^i\}_{i \in S}$ searched for exist and are continuous not only for $t \in (0, T)$, but also for $t \in (0, T)$. It is so, as this fact is used to construct appropriate difference schemes and to prove the consistency and convergence of the numerical method used. Thus, the assumption that a solution of problem (2.2), (2.7), and (2.8) is a $C_S^{2+\alpha}(\overline{D})$ function is not sufficient. One has to consider functions continuous in $\overline{D}$, with continuous derivatives $\mathcal{D}_t w^i$, $\mathcal{D}_{x_j} w^i$, and $\mathcal{D}^2_{x_j x_k} w^i$ ($j, k = 1, \ldots, m; i \in S$) in $\overline{D}$. This means one has to consider the Hölder spaces $H^{k+\alpha, \frac{k+\alpha}{2}}(\overline{D})$ in Ladyzhenskaya's sense. Therefore, we adopt the following definitions.

The *Hölder space* $H^{k+\alpha, \frac{k+\alpha}{2}}(\overline{D})$ ($k = 0, 1, 2; 0 < \alpha < 1$) is a space of functions $w = w(t, x)$ that are continuous in $\overline{D}$, together with all derivatives of the form $\mathcal{D}_t^r \mathcal{D}_x^s w(t, x)$ for $0 \leq 2r + s \leq k$, and with the finite norm

$$|w|^{k+\alpha} := \langle w \rangle^{k+\alpha} + \sum_{l=1}^{k} \langle w \rangle^l,$$

where the components $\langle w \rangle^{k+\alpha}$ and $\langle w \rangle^l$ are given. This is a Banach space.

By $H_S^{k+\alpha, \frac{k+\alpha}{2}}(\overline{D})$ we denote a space of functions $w = \{w^i\}_{i \in S}$ such that $w^i \in H^{k+\alpha, \frac{k+\alpha}{2}}(\overline{D})$ for all $i \in S$, with the finite norm

$$\|w\|^{k+\alpha} := \sup \left\{ |w^i|^{k+\alpha} : i \in S \right\}.$$

Remark 5.1 There is

$$H^{0+\alpha, 0+\frac{\alpha}{2}}(\overline{D}) := H^{\alpha, \frac{\alpha}{2}}(\overline{D}) = C^{0+\alpha}(\overline{D}),$$
$$H^{2+\alpha, 1+\frac{\alpha}{2}}(\overline{D}) \subset C^{2+\alpha}(\overline{D}),$$

where $C^{k+\alpha}(\overline{D})$ stands for the Hölder space in Friedman's sense.

Mathematical Neuroscience. http://dx.doi.org/10.1016/B978-0-12-411468-5.00005-3

A *function* $w = \{w^i\}_{i \in S} \in C_S(\overline{D})$ will be called *∗-regular* in $\overline{D}$ if the functions w^i, $i \in S$, have continuous derivatives $\mathcal{D}_t w^i$, $\mathcal{D}_{x_j} w^i$, and $\mathcal{D}^2_{x_j x_k} w^i$ in $\overline{D}$ for $j, k = 1, \ldots, m$.

We consider system (4.1) with initial and boundary conditions (2.7) and (2.8), i.e., the problem

$$\begin{cases} \mathcal{F}^i[z^i](t, x) = f^i(t, x, z(t, x), z), & \text{for } (t, x) \in D, \\ z^i(0, x) = \phi^i_0(x) & \text{for } x \in G, \\ z^i(t, x) = h^i(t, x) & \text{for } (t, x) \in \sigma \quad \text{and} \quad i \in S, \end{cases} \tag{5.1}$$

where the operators $\mathcal{F}^i$, $i \in S$, are uniformly parabolic in $\overline{D}$, and f^i, $i \in S$, are given functions

$$f^i : \overline{D} \times \mathcal{B}(S) \times C_S(\overline{D}) \to \mathbb{R}, \quad (t, x, y, s) \mapsto f^i(t, x, y, s).$$

Moreover, we assume that the functions f, ϕ_0, and h which appear in (5.1) satisfy the suitable compatibility conditions. The compatibility conditions for these functions state that the derivatives $\mathcal{D}_t z^i \mid_{t=0}$, $i \in S$, which are determined for $t = 0$ by means of the equations of the system and of the initial condition, also satisfy boundary condition for $x \in \partial G$. Therefore, in our case these functions must satisfy the compatibility conditions of order $\left[\frac{\alpha}{2}\right] + 1$ of the form

$$h(0, x) = \phi_0(x), \quad \mathcal{D}_t h^i(t, x) \mid_{t=0} - \mathcal{L}^i[\phi^i_0](x) = f^i(0, x, \phi_0(x), \phi_0), \tag{5.2}$$

$i \in S$, for $x \in \partial G$.

We assume that $\partial G \in H^{2+\alpha}$ and adopt the following assumptions:

Assumption $\overline{H}_a$. The coefficients $a^i_{jk} = a^i_{jk}(t, x)$, $a^i_{jk} = a^i_{kj}$, and $b^i_j = b^i_j(t, x)$ $(j, k = 1, \ldots, m, i \in S)$ of operators $\mathcal{L}^i$, $i \in S$, are locally Hölder continuous with exponent α $(0 < \alpha < 1)$ with respect to t and x in $\overline{D}$ and their Hölder norms are uniformly bounded.

Assumption $\overline{H}_f$. $f(\cdot, \cdot, y, s) \in H_S^{\alpha, \frac{\alpha}{2}}(\overline{D})$, where $0 < \alpha < 1$.

Assumption $\overline{H}_{\phi, h}$. $\phi_0 \in H_S^{2+\alpha}(\overline{G})$, $h \in H_S^{2+\alpha, 1+\frac{\alpha}{2}}(\sigma)$, where $0 < \alpha < 1$.

Assumption $\overline{A}_0$. We assume that there exists at least one ordered pair u_0 and v_0 of a lower and an upper solution of problem (5.1) in $\overline{D}$, and $u_0, v_0 \in H_S^{\alpha, \frac{\alpha}{2}}(\overline{D})$.

Let $\beta \in C_S(\overline{D})$ be a sufficiently regular function. Denote by $\mathcal{P}$ the operator

$$\mathcal{P} : \beta \mapsto \mathcal{P}[\beta] = \gamma,$$

where γ is the (supposedly unique) solution of the linear initial-boundary value problem

$$\begin{cases} \mathcal{F}^i[\gamma^i](t, x) = f^i(t, x, \beta(t, x), \beta) & \text{for } (t, x) \in D, \\ \gamma^i(0, x) = \phi^i_0(x) & \text{for } x \in G, \\ \gamma^i(t, x) = h^i(t, x) & \text{for } (t, x) \in \sigma \quad \text{and} \quad i \in S, \end{cases}$$

with compatibility conditions (5.2).

The operator $\mathcal{P}$ is the composition of the nonlinear operator $\mathbf{F} = \{\mathbf{F}^i\}_{i \in S}$ generated by the functions $f^i(t, x, y, s), i \in S$, and defined for any $\beta \in C_S(\overline{D})$ as follows

$$\mathbf{F}: \beta \mapsto \mathbf{F}[\beta] = \delta,$$

where

$$\mathbf{F}^i[\beta](t, x) := f^i(t, x, \beta(t, x), \beta) = \delta^i(t, x), \quad i \in S,$$

and the operator

$$\mathcal{G}: \delta \mapsto \mathcal{G}[\delta] = \gamma,$$

where γ is the (supposedly unique) solution of the linear initial-boundary value problem

$$\begin{cases} \mathcal{F}^i[\gamma^i](t, x) = \delta^i(t, x), & \text{for } (t, x) \in D, \\ \gamma^i(0, x) = \phi_0^i(x) & \text{for } x \in G, \\ \gamma^i(t, x) = h^i(t, x) & \text{for } (t, x) \in \sigma, \quad \text{and } i \in S, \end{cases} \tag{5.3}$$

with compatibility conditions (5.2).

Hence

$$\mathcal{P} = \mathcal{G} \circ \mathbf{F}.$$

Using a similar argument as in Section 4.1, the proof of the following auxiliary lemmas and the theorem readily follow.

Lemma 5.1 *If $\beta \in H_S^{\alpha, \frac{\alpha}{2}}(\overline{D})$ and the function $f = \{f^i\}_{i \in S}$ generating the nonlinear operator $\mathbf{F}$ satisfies assumptions $(\overline{\mathbf{H}}_{\mathbf{f}})$ and $(\mathbf{L})$, then*

$$\mathbf{F}[\beta] = \delta \in H_S^{\alpha, \frac{\alpha}{2}}(\overline{D}).$$

Lemma 5.2 *If $\delta \in H_S^{\alpha, \frac{\alpha}{2}}(\overline{D})$ and assumptions $(\overline{\mathbf{H}}_{\mathbf{a}})$, $(\overline{\mathbf{H}}_{\phi, \mathbf{h}})$ hold and $\partial G \in H^{2+\alpha}$, then the problem (5.3) has the unique $*$-regular solution γ, and $\gamma \in H_S^{2+\alpha, 1+\frac{\alpha}{2}}(\overline{D})$.*

Corollary 5.1 *If the assumptions of Lemmas 5.1 and 5.2 hold, then the operator $\mathcal{P}$ has the following property:*

$$\mathcal{P}: H_S^{\alpha, \frac{\alpha}{2}}(\overline{D}) \to H_S^{2+\alpha, 1+\frac{\alpha}{2}}(\overline{D}).$$

Theorem 5.1 *Let assumptions $\overline{\mathbf{A}_0}$, $(\overline{\mathbf{H}}_{\mathbf{a}})$, $(\overline{\mathbf{H}}_{\mathbf{f}})$, $(\overline{\mathbf{H}}_{\phi, \mathbf{h}})$ hold, and $(\mathbf{W})$, $(\mathbf{L})$, $(\mathbf{V})$ hold in the set $\mathcal{K}$. If the successive terms of the approximation sequences $\{u_n\}$ and $\{v_n\}$ are defined as $*$-regular solutions of the following infinite systems of equations:*

$$\begin{cases} \mathcal{F}^i[u_n^i](t, x) = f^i(t, x, u_{n-1}(t, x), u_{n-1}), & \text{for } (t, x) \in D, \\ u_n^i(0, x) = \phi_0^i(x) & \text{for } x \in G, \\ u_n^i(t, x) = h^i(t, x) & \text{for } (t, x) \in \sigma \quad \text{and } i \in S \end{cases}$$

and

$$\begin{cases} \mathcal{F}^i[v_n^i](t, x) = f^i(t, x, v_{n-1}(t, x), v_{n-1}), & \text{for } (t, x) \in D, \\ v_n^i(0, x) = \phi_0^i(x) & \text{for } x \in G, \\ v_n^i(t, x) = h^i(t, x) & \text{for } (t, x) \in \sigma \text{ and } i \in S, \end{cases}$$

(i.e., if we use the method of direct iterations), where an ordered pair of a lower u_0 and an upper v_0 solution will be the pair of initiated iteration in this iterative process, then there exists the unique global $*$-regular solution z of problem (5.1) within the sector $\langle u_0, v_0 \rangle$, and $z \in H_S^{2+\alpha, 1+\frac{\alpha}{2}}(\overline{D})$.

5.2 On Constructions of Upper and Lower Solutions

5.2.1 Fundamental Problem

It is well known that the method of upper and lower solutions may be used to prove the existence of solutions of many types of differential equations. Applying this method, we also obtain a sector in which those solutions must be considered. This is *a priori* important information about the localization of solutions. A basic difficulty in applying monotone iterative methods lies in the construction of a pair of a lower and an upper solution of the considered problem. The literature on monotone methods describes no general way to build such functions. However, the right-hand sides of the equations discussed, i.e., the functions $f^i(t, x, y, s), i \in S$, should be bounded in the domain considered. A lower and an upper solution can be easily built by means of the Green function for the equation and the domain considered.

Consider an infinite system of equations with homogeneous initial-boundary condition in cylindrical domain D, i.e., the problem

$$\begin{cases} \mathcal{F}^i[z^i](t, x) = f^i(t, x, z(t, x), z), & \text{for } (t, x) \in D \text{ and } i \in S, \\ z(t, x) = 0 & \text{for } (t, x) \in \Gamma \end{cases}$$

and let functions $f^i(t, x, y, s), i \in S$, in the set $\overline{D} \times \mathcal{B}(S) \times C_S(\overline{D})$ and parabolic operators $\mathcal{F}^i, i \in S$, fulfill suitable assumptions. Using the fundamental solutions $\Gamma^i(t, x; \tau, \xi)$ for the equations $\mathcal{F}^i[z^i](t, x) = 0, i \in S$, we define the Green functions $\mathcal{G}^i(t, x; \tau, \xi), i \in S$ for the domain D and we consider the functions

$$U_0^i(t, x) = m_i(2\sqrt{\pi})^{-m} \int_0^t \int_G \mathcal{G}^i(t, x; \tau, \xi) d\tau d\xi, \tag{5.4}$$

$$V_0^i(t, x) = M_i(2\sqrt{\pi})^{-m} \int_0^t \int_G \mathcal{G}^i(t, x; \tau, \xi) d\tau d\xi, \quad i \in S, \tag{5.5}$$

where

$$m_i = \inf_{\overline{D} \times \mathcal{B}(S) \times C_S(\overline{D})} f^i(t, x, y, s),$$

$$M_i = \sup_{\overline{D} \times \mathcal{B}(S) \times C_S(\overline{D})} f^i(t, x, y, s) \quad \text{for } i \in S.$$

These functions fulfill homogeneous initial-boundary condition (2.6) and the equations

$$\mathcal{F}^i[U_0^i](t, x) = m_i,$$
$$\mathcal{F}^i[V_0^i](t, x) = M_i \quad \text{for } (t, x) \in D \quad \text{and} \quad i \in S.$$

Therefore, the following differential inequalities hold.

$$\mathcal{F}^i[U_0^i](t, x) - f^i(t, x, U_0(t, x), U_0) = m_i - f^i(t, x, U_0(t, x), U_0) \le 0,$$
$$\mathcal{F}^i[V_0^i](t, x) - f^i(t, x, V_0(t, x), V_0) = M_i - f^i(t, x, V_0(t, x), V_0) \ge 0,$$
$$\text{for } (t, x) \in D \quad \text{and} \quad i \in S.$$

Consequently, the functions U_0 and V_0 are the lower and the upper solutions of the problem considered in $\overline{D}$, respectively, and one may use them as the pair of initial iterations in the iterative process.

5.2.2 Example

A procedure which makes use of the comparison principle (Theorem 3.1) in the construction of a lower and an upper solution of the following equation:

$$\begin{cases} \frac{\partial \theta}{\partial t} = \frac{\partial^2 \theta}{\partial x^2} + d \exp\left(\frac{\alpha \theta}{\alpha + \theta}\right) & \text{for } (t, x) \in D, \\ \theta(0, x) = h(x) & \text{for } (t, x) \in G, \\ \theta(t, x) = 0 & \text{for } (t, x) \in \sigma, \end{cases}$$

where $\theta = \theta(t, x)$, d, and α are positive parameters and $D = (0, T) \times G$, where G is a long cylinder.

In the case $0 < x < 1$, the steady-state lower and upper solutions are sought in the form

$$u_0(x) = C \sin^2 \pi x$$
$$v_0(x) = 4kx(1 - x),$$

where $C, k > 0$ are constants to be determined.

The solution θ of the problem considered is such that

$$C \sin^2 \pi x \le \lim_{t \to +\infty} \theta(t, x) \le 4kx(1 - x)$$

for some constants C and k.

5.3 Positive Solutions

For the purposes of applications, an important task is to find *positive* (nonnegative) *solutions*. In general, this is a difficult problem.

Let us consider problem (2.2), (2.7), and (2.8) in $\overline{D}$, i.e.,

$$
\begin{cases}
\mathcal{D}_t z^i(t, x) - \mathcal{L}^i[z^i](t, x) = f^i(t, x, z) & \text{for } (t, x) \in D, \\
z^i(0, x) = \phi_0^i(x) & \text{for } x \in G, \\
z^i(t, x) = h^i(t, x) & \text{for } (t, x) \in \sigma \quad \text{and} \quad i \in S
\end{cases}
\tag{5.6}
$$

with compatibility condition

$$
h^i(0, x) = \phi_0^i(x) \quad \text{for } x \in \partial G,
$$

where

$$
\mathcal{L}^i = \sum_{j,k=1}^{m} a_{jk}^i(t, x)\mathcal{D}_{x_j x_k}^2 - \sum_{j=1}^{m} b_j^i(t, x)\mathcal{D}_{x_j}, \quad i \in S.
$$

If $u = u(t, x) = 0$ in $\overline{D}$ is the lower solution of problem (5.6) in $\overline{D}$, then the following inequalities:

$$
\begin{aligned}
f^i(t, x, 0) &\geq 0, \quad \text{in } D, \\
\phi_0^i(x) &\geq 0 \quad \text{on } G, \\
h^i(t, x) &\geq 0 \quad \text{on } \sigma \quad \text{and} \quad i \in S,
\end{aligned}
\tag{5.7}
$$

hold.

The following theorem may be directly proved.

Theorem 5.2   *Let us consider problem* (5.6) *and let:*

   i.   *the inequalities hold and not all the three functions be identically zero simultaneously;*
   ii. *there exists a positive upper solution v_0 of this problem in $\overline{D}$;*
   iii. *the functions $f^i(t, x, s), i \in S$, satisfy the conditions* (**W**), (**V**) *and the left-hand side Lipschitz condition* (**L₁**) *with respect to s, for $s \in \langle 0, v_0 \rangle$.*

Then problem (5.6) *has at last one positive solution z within the sector $\langle 0, v_0 \rangle$.*

Moreover, if $f^i(t, x, 0) = \phi_0(x) = \psi(t, x) = 0$, then this solution $z = z(t, x) \equiv 0$ in $\overline{D}$.

This means that the application of the existence theorem in the case of a positive solution requires a positive upper solution v_0 to be constructed, which may prove difficult.

5.4 Some Remarks on Strongly Coupled Systems

We have confined ourselves to studying weakly coupled systems of second-order partial differential equations. In the case of *strongly coupled*[1] systems of first-order partial differential equations, the Cauchy problem is not well posed. This means that the situation gets intrinsically complex.

There is the well-known fundamental example of the nonuniqueness of the Cauchy problem for the strongly coupled system of first-order equations of the form

$$\frac{\partial z_j}{\partial x} = \sum_{k=1}^{2} a_{jk}(x, y) \frac{\partial z_k}{\partial y}, \quad j = 1, 2 \quad \text{for } (x, y) \in \mathbb{R}^2,$$

with the initial condition

$$z_j(0, y) = 0, \quad j = 1, 2 \quad \text{for } y \in \mathbb{R},$$

where the coefficients a_{jk} ($j, k = 1, 2$) are defined in the whole plane $\mathbb{R}^2$ and $a_{jk} \in C^\infty(\mathbb{R}^2)$. This problem does not have a unique solution in the class $C^\infty(\mathbb{R}^2)$. Precisely, this problem has a solution of the class $C^\infty(\mathbb{R}^2)$, vanishing together with all its derivatives for $x = 0$, and not vanishing identically in any neighborhood of the point $(0, 0)$.

This new approach to the study of strongly coupled systems has been studied by numerous authors. However, this method is only appropriate for linear systems of equations whose coefficients are independent of a so-called Bernstein variable.

5.5 Estimation of Convergence Speed for Different Iterative Methods

Let $\{u_n\}$ be a sequence of successive approximations which converges to a solution z of problem (4.1)

$$z = z(t, x) = \lim_{n \to \infty} u_n(t, x) \quad \text{for } (t, x) \in \overline{D}.$$

We shall say that the sequence $\{u_n\}$ converges to the solution z with a *geometrical speed*, if the following inequalities hold:

$$\|u_n - z\|_0 \le cq^n, \quad n = 1, 2, \ldots, \quad \text{in } \overline{D}, \tag{5.8}$$

where $c = \text{const} > 0$ and $0 < q < 1$.

If

$$\|u_n - z\|_0 \le c \frac{K^n}{n!}, \quad n = 1, 2, \ldots, \quad \text{in } \overline{D}, \tag{5.9}$$

---

[1] The strong coupling of a system of second-order partial differential equations means that each first-order spatial derivative may appear in each equation of this system.

where c and K are nonnegative constants, then we say that $\{u_n\}$ converges to z with a *power speed*.

The convergence in this sense is essentially faster than the convergence with a geometrical speed.

We shall say that $\{u_n\}$ converges to z with a *Newton speed*, if

$$\|u_n - z\|_0 \leq c\delta^{2^n}, \quad \text{for } (t, x) \in D, \quad i \in S \text{ and } n = 1, 2, \ldots, \tag{5.10}$$

where $c = \text{const} > 0$ and $0 < \delta < 1$. This convergence is essentially faster than the previous one.

If the successive terms of sequences $\{u_n\}$ and $\{v_n\}$ are lower and upper solutions of problem (4.1), defined, respectively, as solutions of systems (4.2) and (4.3), where an initial iteration is given by the ordered pair u_0 and v_0 given by assumption A_0, then estimates (4.7) hold

$$0 \leq v_n^i(t, x) - u_n^i(t, x) \leq N_0 \frac{[(L_1 + L_2)t]^n}{n!}, \quad n = 0, 1, 2, \ldots, \quad i \in S,$$

for $(t, x) \in \overline{D}$, where $N_0 = \|v_0 - u_0\|_0 < \infty$.

Therefore, these sequences converge to the exact solution with a power speed.

We have almost the same as for Chaplygin's sequences $\{\hat{u}_n\}$ and $\{\hat{v}_n\}$ defined as solutions of (4.25) and (4.26) where an initial iteration is given by the ordered pair u_0 and v_0 given by assumption A_0. Estimate (4.29) proves that these sequences converge with at least power speed.

Analogously, by (4.49), the sequences $\{\tilde{u}_n\}$ and $\{\tilde{v}_n\}$, defined as solutions of systems (4.46) and (4.47), converge to z with at least power speed.

Chaplygin's method applied to the Fourier first initial-boundary value problem of the form

$$\begin{cases} \dfrac{\partial z}{\partial t} - \dfrac{\partial^2 z}{\partial x^2} = f(t, x, z(t, x)) & \text{for } (t, x) \in (0, T) \times (a, b) := D, \\ z(t, x) = \psi(t, x) & \text{for } (t, x) \in \Gamma. \end{cases} \tag{5.11}$$

assumed that there existed a lower u_0 and an upper v_0 solution, and defined a sequence of lower function $\{u_n\}$ as regular solutions of the following equations:

$$\frac{\partial u_n}{\partial t} - \frac{\partial^2 u_n}{\partial x^2} = f(t, x, u_{n-1}(t, x)) + f_y(t, x, u_{n-1}(t, x)) \cdot [u_n(t, x) - u_{n-1}(t, x)], \tag{5.12}$$

$$\text{for } (t, x) \in D \text{ and } n = 1, 2, \ldots.$$

Next the convergence of the sequence $\{u_n\}$ to the exact solution of this problem, assuming additionally that the function $f(t, x, y)$ is sufficiently regular and the derivative $f_y(t, x, y)$

satisfies the Lipschitz condition with respect to y. In particular, assuming that

$$\sup_{(t,x,y)\in\mathcal{K}^*} |f_y(t, x, y)| = c_0 < +\infty,$$

$$\sup_{(t,x,y)\in\mathcal{K}^*} |f_{yy}(t, x, y)| = H < +\infty$$

and $f_{yy} > 0$, the estimates follow:

$$|u_n(t, x) - z(t, x)| \leq \frac{2C}{2^{2^n}}, \quad \text{for } (t, x) \in \overline{D} \text{ and } n = 1, 2, \ldots,$$

where

$$C = \frac{1}{2HTe^{c_0 T}}.$$

This means that the convergence speed is quadratic and the sequence of successive approximations converges uniformly to the unique solution with the Newton speed.

We cannot directly repeat the above results on the speed of convergence of successive approximations in the case of equation of the form

$$\frac{\partial z}{\partial t} - \frac{\partial^2 z}{\partial x^2} = f(t, x, z(t, x), z). \tag{5.13}$$

This is so because, even in the simple case considered here, the algorithms have not guaranteed convergence with this speed. To prove results similar to those above we need stronger assumptions on the regularity of function $f(t, x, y, s)$, or we will have to use the full quasilinearization of (5.13) with respect to both arguments y and s, simultaneously. This means that we need to define the sequence of successive approximations $\{u_n^*\}$ as a sequence of solutions of the following equations:

$$\begin{aligned}
\frac{\partial u_n^*(t, x)}{\partial t} - \frac{\partial^2 u_n^*(t, x)}{\partial x^2} &= f(t, x, u_{n-1}^*(t, x), u_{n-1}^*) \\
&\quad + f_y(t, x, u_{n-1}^*(t, x), u_{n-1}^*) \cdot [u_n^*(t, x) - u_{n-1}^*(t, x)] \\
&\quad + f_s(t, x, u_{n-1}^*(t, x), u_{n-1}^*) \cdot [u_n^* - u_{n-1}^*], \quad \text{for } n = 1, 2, \ldots,
\end{aligned} \tag{5.14}$$

where f_s is the Fréchet derivative of a function $f = f(t, x, y, s)$ with respect to the functional argument s. This convergence speed is quadratic.

CHAPTER 6

Truncation Method

6.1 Introduction

We remark that it is not possible to solve directly infinite countable and uncountable systems of equations. In practice, an infinite system of equations is replaced by a finite system of suitably defined equations. Such a transition from infinite systems of equations to finite ones, known as truncation, may be effected in various ways. One of the ways to describe the truncation process is with a projection from an appropriately chosen infinite-dimensional vector space onto its finite-dimensional subspaces.

In the case of infinite countable systems, this will be the classical partially ordered infinite-dimensional Banach space of convergent sequences of real-valued functions $\mathscr{C}_{\mathbb{N}}(\overline{D})$ and then $C_N(\overline{D})$ will be its finite-dimensional subspace. In the case of infinite uncountable systems the studies are carried out in the Banach space L_1 which is natural from the physical point of view or in a suitable chosen Hilbert space.[1]

In the truncation method, solutions of the infinite systems of equations are defined as the limits when $N \to \infty$ of the sequences of approximations which are solutions of the finite truncated systems of the first N equations in N unknown functions with the corresponding initial-boundary conditions. However, we observe that we do not need to know the previous approximations to determine the next approximations.

In the second step of the truncation method one has to prove that finite truncated systems of considered equations have solutions in the suitable space.

It should be stressed here that there are many existence theorems for finite systems of equations. These existence theorems have been proved by means of the monotone iterative method (method of lower and upper solutions), the topological fixed point method, the theory of continuous semigroups of linear operators and evolution systems techniques, as well as the finite difference method, to mention just a few most commonly used.

---

[1] Hilbert spaces play an important role in quantum cognition.

Mathematical Neuroscience. http://dx.doi.org/10.1016/B978-0-12-411468-5.00006-5

If the truncated systems are solved by the method of continuous semigroups of linear operators, then we must recast the considered system with the initial condition as an abstract Cauchy problem. Next, by using the truncation argument and applying a certain fixed point theorem, we prove that, under some assumptions, the abstract Cauchy problem has a unique, global in time solution.

If the truncated systems of equations are solved by means of the finite-difference method, then we will have the three basic computational monotone iterative schemes to choose from: the Picard iteration, a modified version of Jacobi iteration, or the Gauss-Seidel method. If the initial iteration is always a pair of known coupled lower and upper solutions of the problem considered, then the sequence generated by the Picard iteration converges faster than the sequence generated by the Gauss-Seidel iteration, which, in turn, converges faster than the sequence generated by the Jacobi iteration.

We remark that the finite-difference method is not only one of the simplest methods used in numerical analysis, but is also an important analytical method of proving existence theorems in the field of partial differential equations. It is highly advantageous to choose a finite-difference method to prove the existence and uniqueness of any truncated system for the problem considered, because we thus arrive at numerically proved constructive theorems on existence. Each of these solutions is an approximation of a solution of this problem and may forthwith be calculated and plotted or tabulated.

Applying the method of lower and upper solutions requires assuming the monotonicity and the Lipschitz condition of the reaction functions with respect to the function and functional arguments y and s, respectively. The right-hand side Lipschitz condition is used to ensure the uniqueness of the solution (it follows from Szarski's uniqueness criterion) and the left-hand side condition is necessary to ensure the existence of the solution. We also assume the existence of an ordered pair of lower and upper solutions of the problem considered.

The third step of the truncation method will be finding an additionally sufficient condition, which, while added to the previous assumptions, will guarantee the existence of a regular limit function for approximating sequences of solutions of the truncated systems, where this function—in line with the definition adopted—will be a solution to our problem. The condition requiring the reaction functions to be bounded by the terms of a convergent sequence turns out to be such a sufficient condition. Therefore, we assume the following sufficient condition (B):

Suppose that the reaction functions $f^j = f^j(t, x, y, s)$, $j \in \mathbb{N}$, are continuous and there exists a sequence $\{q_j\}_{j \in \mathbb{N}}$ of nonnegative real numbers $q_j \geq 0$ for $j \in \mathbb{N}$ such that $|f^j(t, x, y, s)| \leq q_j$ for $j \in \mathbb{N}$, where $\lim_{j \to \infty} q_j = 0$.

This condition plays a crucial role in proving the existence theorem. Similar conditions occur where infinite countable systems are studied.

We adopt the notations of Chapter 2 and introduce the convention that every infinite sequence

$$w_{N,0} = (w_N^1, \ldots, w_N^j, \ldots, w_N^N, 0, 0, \ldots) \in \mathscr{C}_{N,0}(\overline{D})$$

is treated as a finite one

$$w_N = (w_N^1, \ldots, w_N^j, \ldots, w_N^N) \in C_N(\overline{D}),$$

which we will write as

$$w_{N,0} \cong w_N \quad \text{for all } N \in \mathbb{N}.$$

In this sense, the space $\mathscr{C}_{N,0}(\overline{D})$ is identified with the space $C_N(\overline{D})$, which we will write as

$$\mathscr{C}_{N,0}(\overline{D}) \cong C_N(\overline{D}). \tag{6.1}$$

Finally, in this sense, the space $C_N(\overline{D})$ may be treated as the subspace of the space $\mathscr{C}_\mathbb{N}(\overline{D})$.

6.2 Truncation Method for Infinite Countable Systems

Let us consider weakly coupled[2] infinite countable systems (shortly: countable systems) of equations of the form[3]

$$\mathcal{F}^j[z^j](t, x) := \mathcal{D}_t z^j(t, x) - \mathcal{L}^j[z^j](t, x) = f^j(t, x, z(t, x), z) \tag{6.2}$$

for $j \in \mathbb{N}$, where

$$\mathcal{L}^j := \sum_{l,k=1}^m a_{lk}^j(t, x)\mathcal{D}_{x_l x_k}^2 - \sum_{i=1}^m c_i^j(t, x)\mathcal{D}_{x_i}$$

are diffusion-convection operators and $x = (x_1, \ldots, x_m)$, $(t, x) \in (0, T] \times G := D$, $0 < T < \infty$, where T can be arbitrarily large, $G \subset \mathbb{R}^m$ and G is an open and bounded domain, whose boundary ∂G is an $(m - 1)$-dimensional surface of a class $C^{2+\alpha}$ $(0 < \alpha < 1)$, $S_0 := \{(t, x) : t = 0, x \in \overline{G}\}$, $\sigma := [0, T] \times \partial G$ is a lateral surface of a cylindrical domain D, $\Gamma := S_0 \cup \sigma$ is the parabolic boundary of domain D and $\overline{D} := D \cup \Gamma$, $\mathbb{N}$ is the set of natural numbers and N is an arbitrary fixed natural number.

Diagonal operators $\mathcal{F}^j$, $j \in \mathbb{N}$, are uniformly parabolic in $\overline{D}$, z stands for the functions

$$z: \mathbb{N} \times \overline{D} \to \mathbb{R}, \quad (j, t, x) \mapsto z(j, t, x) := z^j(t, x),$$

---

[2] This means that every equation contains derivatives of one unknown function only.

[3] The notation w denotes that w is regarded as an element of the set of admissible functions, while $w(t, x)$ stands for the value of this function at time t and on the point x. However, sometimes, to stress the dependence of function w on the variables t and x, we will write $w = w(t, x)$ and hope that this will not confuse the reader.

composed of unknown functions $z = \{z^j\}_{j \in \mathbb{N}} := (z^1, z^2, \ldots)$, and f^j, $j \in \mathbb{N}$, are given nonlinear functions

$$f^j : \Omega := \overline{D} \times \ell^\infty \times \mathscr{C}_\mathbb{N}(\overline{D}) \to \mathbb{R}, \quad (t, x, y, s) \mapsto f^j(t, x, y, s), \quad j \in \mathbb{N}.$$

The right-hand sides f^j of the equations, i.e., the reaction functions (reaction terms) which describe kinetic behavior of considered process, are functionals with respect to the last variable and we assume that they are of the Volterra type.

If we introduce the function $\tilde{f} = \{\tilde{f}^j\}_{j \in \mathbb{N}}$ setting

$$\tilde{f}^j(t, x, z) := f^j(t, x, z(t, x), z), \quad j \in \mathbb{N},$$

where

$$\tilde{f}^j : \tilde{\Omega} := \overline{D} \times \mathscr{C}_\mathbb{N}(\overline{D}) \to \mathbb{R}, \quad (t, x, s) \mapsto \tilde{f}^j(t, x, s), \quad j \in \mathbb{N},$$

then we will write the equations of system (6.2) in another form, which may be useful in our further considerations:

$$\mathcal{F}^j[z^j](t, x) = \tilde{f}^j(t, x, z) := \tilde{f}^j(t, x, z^1, z^2, \ldots) \tag{6.3}$$

for $j \in \mathbb{N}$, $(t, x) \in D$.

For system (6.2) we will consider the Fourier first initial-boundary value problem: find the regular (classical) solution of system (6. 2) (or (6. 3)) in $\overline{D}$ fulfilling the initial-boundary condition

$$z(t, x) = \phi(t, x) \quad \text{for } (t, x) \in \Gamma \tag{6.4}$$

or the homogeneous initial-boundary condition

$$z(t, x) = 0 \quad \text{for } (t, x) \in \Gamma. \tag{6.5}$$

In the truncation method a solution z of infinite countable system (6.3) is defined as the limit when $N \to \infty$ of the sequences of approximations $\{z_N\}_{N=1,2,\ldots}$, where $z_N = (z_N^1, z_N^2, \ldots, z_N^N)$ are defined as solutions of finite systems of the first N equations of system (6.3) in N unknown functions (i.e., truncated system) of the form

$$\mathcal{F}^j[z_N^j](t, x) = \tilde{f}^j\left(t, x, z_N^1, \ldots, z_N^j, \ldots, z_N^N, \psi^{N+1}, \psi^{N+2}, \ldots\right)$$
$$:= \tilde{f}_{N,\psi}^j(t, x, z_N) \quad \text{for } j = 1, 2, \ldots, N, \quad (t, x) \in D \tag{6.6}$$

with the corresponding initial-boundary conditions of the form

$$z_N^j(t, x) = \phi^j(t, x) \quad \text{for } j = 1, 2, \ldots, N, \quad (t, x) \in \Gamma. \tag{6.7}$$

The remaining terms $z_N^{N+1}, z_N^{N+2}, \ldots$ of the approximation sequences $\{z_N\}_{N=1,2,\ldots}$ are defined as follows:

$$z_N^j(t, x) := \psi^j(t, x) \quad \text{for } j = N + 1, N + 2, \ldots, \quad (t, x) \in \overline{D}, \tag{6.8}$$

where the function $\psi = \{\psi^j\}_{j \in \mathbb{N}}$, $\psi^j = \psi^j(t, x)$, defined for $(t, x) \in \overline{D}$, and satisfying initial-boundary condition (6.7)

$$\psi(t, x) = \phi(t, x) \quad \text{for } (t, x) \in \Gamma,$$

will be determined later on.

Let us consider the infinite-dimensional Banach sequence space $\mathscr{C}_{\mathbb{N}}(\overline{D})$ and this subspace $\mathscr{C}_{N,0}(\overline{D})$.

We define the operator ρ_N as follows:

$$\rho_N : \quad \mathscr{C}_{\mathbb{N}}(\overline{D}) \to \mathscr{C}_{N,0}(\overline{D}), \tag{6.9}$$

$$(z^1, z^2, \ldots) = z \mapsto \rho_N[z] := z_{N,0} = (z^1, \ldots, z^j, \ldots, z^N, 0, 0, \ldots)$$

for all $z \in \mathscr{C}_{\mathbb{N}}(\overline{D})$ and an arbitrary $N \in \mathbb{N}$.

This means that

$$\rho_N[z] := z_{N,0} = \begin{cases} z^j & \text{when } 1 \le j \le N, \\ 0 & \text{when } j > N \end{cases} \tag{6.10}$$

for all $z \in \mathscr{C}_{\mathbb{N}}(\overline{D})$ and an arbitrary $N \in \mathbb{N}$.

It is easy to see that

$$\rho_N^2[z] = \rho_N[\rho_N[z]] = \rho_N[z_{N,0}] = z_{N,0} = \rho_N[z]$$

for all $z \in \mathscr{C}_{\mathbb{N}}(\overline{D})$ and $N \in \mathbb{N}$. Therefore, ρ_N is the projection operator.

We define also the projection operator q_N by:

$$q_N : \quad \mathscr{C}_{\mathbb{N}}(\overline{D}) \to C_N(\overline{D}), \tag{6.11}$$

$$(z^1, z^2, \ldots) = z \mapsto q_N[z] := z_N = (z^1, z^2, \ldots, z^N)$$

for all $z \in \mathscr{C}_{\mathbb{N}}(\overline{D})$ and an arbitrary $N \in \mathbb{N}$.

By convention (6.1), we treat that

$$\rho_N[z] \cong q_N[z] \tag{6.12}$$

for all $z \in \mathscr{C}_{\mathbb{N}}(\overline{D})$ and $N \in \mathbb{N}$.

Example Let us consider the infinite countable system of equations of the form (6.2) with initial-boundary condition (6.4), i.e.,

$$
\begin{cases}
\mathcal{F}^j[z^j](t, x) = f^j(t, x, z(t, x), z) := \tilde{f}^j(t, x, z) & \text{for } (t, x) \in D, \\
z^j(t, x) = \phi^j(t, x) & \text{for } (t, x) \in \Gamma,
\end{cases}
\tag{6.13}
$$

for $j \in \mathbb{N}$, in the infinite-dimensional Banach sequence space $\mathscr{C}_{\mathbb{N}}(\overline{D})$, where the reaction functions have the special form:

$$
\tilde{f}^1(t, x, z) := -z^1(t, x) \sum_{k=1}^{\infty} a_k^1 z^k(t, x) + \sum_{k=1}^{\infty} b_k^1 z^{1+k}(t, x),
$$

$$
\tilde{f}^j(t, x, z) := \frac{1}{2} \sum_{k=1}^{j-1} a_k^{j-k} z^{j-k}(t, x) z^k(t, x)
$$

$$
- z^j(t, x) \sum_{k=1}^{\infty} a_k^j z^k(t, x)
\tag{6.14}
$$

$$
+ \sum_{k=1}^{\infty} b_k^j z^{j+k}(t, x) - \frac{1}{2} z^j(t, x) \sum_{k=1}^{j-1} b_k^{j-k} \quad \text{for } j = 2, 3 \ldots,
$$

where the coagulation rates a_k^j and the fragmentation rates b_k^j are nonnegative constants such that $a_k^j = a_j^k$ and $b_k^j = b_j^k$.

This system may be treated as the discrete mathematical model of the coagulation-fragmentation processes with diffusion built from infinite countable systems of equations. In the case of solving countable systems of equations as the discrete coagulation-fragmentation models with diffusion, the technique used starts with a study of finite systems obtained by truncation to the first N equations followed by passing to the limit as $N \to \infty$.

If we use the projection operator q_N defined by (6.11) in this problem, then we obtain the finite truncated problem considered in the subspace $C_N(\overline{D})$

$$
\begin{cases}
\mathcal{F}^j[z_N^j](t, x) = \tilde{f}_N^j(t, x, z_N) & \text{for } (t, x) \in D, \\
z_N^j(t, x) = \phi^j(t, x) & \text{for } (t, x) \in \Gamma
\end{cases}
\tag{6.15}
$$

for $j = 1, 2, \ldots, N$ and an arbitrary $N \in \mathbb{N}$, where $z_N = \left(z_N^1, \ldots, z_N^j, \ldots, z_N^N \right) \in C_N(\overline{D})$ and

$$
\tilde{f}_N^1(t, x, z_N) := -z_N^1(t, x) \sum_{k=1}^{N} a_k^1 z_N^k(t, x) + \sum_{k=1}^{N} b_k^1 z_N^{1+k}(t, x),
$$

$$\tilde{f}_N^j(t, x, z_N) = \frac{1}{2} \sum_{k=1}^{j-1} a_k^{j-k} z_N^{j-k}(t, x) z_N^k(t, x) - z_N^j(t, x) \sum_{k=1}^{N} a_k^j z_N^k(t, x)$$

$$+ \sum_{k=1}^{N} b_k^j z_N^{j+k}(t, x) \tag{6.16}$$

$$- \frac{1}{2} z_N^j(t, x) \sum_{k=1}^{j-1} b_k^{j-k} \quad \text{for } j = 2, 3, \ldots, N$$

with the corresponding initial-boundary condition (6.4).

If we truncate this system by acceptance

$$a_k^j \equiv 0 \quad \text{and} \quad b_k^j \equiv 0 \quad \text{for } j > N \quad \text{or} \quad k > N, \tag{6.17}$$

then we obtain the truncated system (6.15) and (6.16).

This means that the truncation of the infinite system to the system of the first N equations of the infinite system in N unknown functions may be treated as a projection of the infinite countable system considered in the infinite-dimensional Banach sequence space $\mathscr{C}_N(\overline{D})$ onto its finite N-dimensional subspace $C_N(\overline{D})$.

6.3 Truncation Method for Infinite Uncountable Systems

The truncation method is also the fundamental approximation method of studying solvability of infinite uncountable systems of ordinary and partial differential equations. A finite truncated system may be obtained from an uncountably infinite system with the use of a projection and investigated by means of the theory of semigroup of linear operators, evolution systems, and fixed point mapping techniques.

Example   Consider the pure multiple-fragmentation process for time-independent kernels $\gamma = \gamma(\lambda, y)$ which is described by the following autonomous integro-differential linear equation:

$$\frac{\partial}{\partial t} z(t, \lambda) = \int_{\lambda}^{\infty} \gamma(y, \lambda) z(t, y) dy - z(t, \lambda) \int_{0}^{\lambda} \frac{y}{\lambda} \gamma(\lambda, y) dy \quad \text{for } \lambda > 0, t > 0 \tag{6.18}$$

with initial condition

$$z(0, \lambda) = f(\lambda) \quad \text{for } \lambda > 0. \tag{6.19}$$

This problem is dependent on an additional real positive parameter[4] $\lambda \in [\lambda_0, \lambda_1] \subset \mathbb{R}^+$ and

$$z : (0, T] \times [\lambda_0, \lambda_1] \to \mathbb{R}^+, \quad (t, \lambda) \mapsto z(t, \lambda). \tag{6.20}$$

We remark that Eq. (6.18) can be treated as a system of uncountable many integro-differential equations.

To apply the theory of semigroups of linear operators we must recast (6.18) with the initial condition (6.19) as a linear abstract Cauchy problem:

$$\begin{cases} \dfrac{d}{dt}\tilde{z}(t) = A[\tilde{z}](t) \quad \text{for } t > 0, \\ \tilde{z}(0) = f, \end{cases} \tag{6.21}$$

where $\tilde{z} = \tilde{z}(t)$ is the suitable defined function.

For this purpose we consider the Banach space $\mathcal{X}$ of type L, precisely the $L_{1,-1}$ space of equivalence classes of measurable, real-valued functions ϕ such that

$$\int_0^\infty \lambda |\phi(\lambda)| d\lambda < \infty$$

equipped with the weighted norm

$$\|\phi\|_{1,-1} := \int_0^\infty \lambda |\phi(\lambda)| d\lambda.$$

Finally, the pure fragmentation equation (6.18) with initial condition (6.19) can be recast as the abstract Cauchy problem

$$\begin{cases} \dfrac{d}{dt} z(t) = A[z](t) \quad \text{for } t > 0, \\ z(0) = f, \end{cases} \tag{6.22}$$

where the linear operator

$$A : L_{1,-1} \supseteq D_{\max}(A) \to L_{1,-1}$$

is defined on its maximal domain $D_{\max}(A)$ by

$$A[\phi](\lambda) := \int_\lambda^\infty \gamma(y, \lambda)\phi(y)dy$$

$$-\phi(\lambda) \int_0^\lambda \frac{y}{\lambda}\gamma(\lambda, y)dy \quad \text{for all } \phi \in D_{\max}(A). \tag{6.23}$$

---

[4] Each cluster is identified by its size which is assumed to be a positive real number. The volume is used as the characteristic size. Therefore, by the size of a cluster we mean its volume.

In this problem the conservation of mass is the identity

$$\int_0^\infty \lambda z(t, \lambda)d\lambda = \int_0^\infty \lambda f(\lambda)d\lambda \quad \text{for all } t \geq 0,$$

which implies that a natural space to work with is a Banach space $\mathcal{X}$ of the type L.

If we consider the subspace L_N of the Banach space $L_{1,-1}$ consisting of functions vanishing when $\lambda \geq N$, $N \in \mathbb{N}$, i.e.,

$$L_N := \{\phi \in L_{1,-1} : \phi = 0 \text{ on } [N, \infty)\}, \tag{6.24}$$

then we define the projection operator P_N on the space $L_{1,-1}$ onto the subspace L_N as follows:

$$P_N : L_{1,-1} \to L_N, \quad z \mapsto P_N[z] := z_N$$

for all $z \in L_{1,-1}$ and an arbitrary $N \in \mathbb{N}$.

This means that

$$P_N[z](\lambda) := \begin{cases} z(\lambda) \text{ when } 0 < \lambda < N, \\ 0 \quad \text{ when } \lambda \geq N \end{cases} \tag{6.25}$$

for all $z \in L_{1,-1}$ and an arbitrary $N \in \mathbb{N}$.

If we use the projection operator P_N defined by (6.25) in the abstract Cauchy problem (6.22), then we obtain the truncated problem

$$\begin{cases} \dfrac{d}{dt}z(t) = A_N[z](t) \quad \text{for } t > 0, \\ z(0) = f, \end{cases} \tag{6.26}$$

where $A_N = A P_N$. The operators A_N are defined as follows:

$$A_N[\phi](\lambda) := \begin{cases} \displaystyle\int_\lambda^N \gamma(y, \lambda)\phi(y)dy - \phi(\lambda)\int_0^\lambda \frac{y}{\lambda}\gamma(\lambda, y)dy \quad \text{for } 0 < \lambda < N, \\ 0 \quad \text{for } \lambda \geq N \end{cases}$$

for all $\phi \in L_{1,-1}$ and an arbitrary $N \in \mathbb{N}$.

Applying the classical methods of the theory of continuous semigroups of linear operators and evolution system techniques and the Banach fixed point theorem for contraction mappings, we prove the existence and uniqueness of a weak solution for the abstract Cauchy problem (6.22).

6.4 Relation Between Continuous and Discrete Infinite-Dimensional Models

Consider the initial value problem of the form

$$\frac{\partial z}{\partial t} - A(t, x, \lambda)z = f(t, x, z(t, x, \lambda), \lambda) \tag{6.27}$$

for $t > 0$, $x \in \mathbb{R}^m$, $m = 1, 2$, or 3, with initial condition

$$z(0, x, \lambda) = z_0(x, \lambda) \quad \text{for } x \in \mathbb{R}^m, \tag{6.28}$$

which is a continuous model of the coagulation-fragmentation processes with diffusion.

This problem depends on an additional real parameter λ, the volume. The diffusion-convection operator A is uniformly elliptic, the reaction function (reaction term) f describes kinetic behavior of the process, and $z = z(t, x, \lambda)$ is the concentration. The above problem can be viewed as an infinite uncountable system of equations.

In the event of the continuous model, the right-hand sides of equations include terms of the form

$$\mathcal{J} = \int_G \int_{\lambda_0}^{\lambda_1} z(t, x, \lambda)dxd\lambda, \tag{6.29}$$

which is the total number of clusters with volumes belonging to the interval $[\lambda_0, \lambda_1] \subset \mathbb{R}^+$ and being at time t contained in the space region $G \subset \mathbb{R}^m$. The measure $d\lambda$ is either Lebesgue's measure on $\mathbb{R}^+$ or the counting measure on $\mathbb{N} := \{1, 2, 3, \ldots\}$. In the latter case only clusters whose sizes are integer multiples of an "elementary unit" can occur. In this case, all integrals with respect to $d\lambda$ reduce to sums, so that (6.29) takes the form:

$$\mathcal{S} = \int_G \sum_{\lambda=\lambda_0}^{\lambda_1} z(t, x, \lambda)dx. \tag{6.30}$$

This situation corresponds to the discrete model of the coagulation-fragmentation processes considered.

Next, consider problem (6.27) and (6.28) as a single semilinear evolution equation

$$\frac{dz}{dt} - A(t)z = f(t, z) \quad \text{for } t > 0 \tag{6.31}$$

with initial condition

$$z(0) = z_0, \tag{6.32}$$

where $z = z(t, x)$ is a Banach-space-valued function of $(t, x) \in \mathbb{R}^+ \times \mathbb{R}^m$. This means that (6.31) may be interpreted as a vector-valued evolution equation which we are able to handle using recent Fourier multiplier theorems for operator-valued symbols and Banach-space-valued distributions.

It should be noted that the main result is the theorem on existence and uniqueness of the solution of the continuous coagulation-fragmentation equation with diffusion in the class of volume preserving solutions. An advantage of this approach is the feasibility of the simultaneous consideration of both the continuous and discrete models of coagulation-fragmentation processes with diffusion.

What is the relationship between solutions of the infinite uncountable system of equations serving as a continuous model of process considered and solutions of the infinite countable system of equations serving as a discrete model of this process? To answer this question let us consider two infinite systems of equations serving as a continuous model (infinite uncountable system) and a discrete model (infinite countable system), respectively, of the same process? To answer this question let us take into consideration expressions of the form (6.29) and (6.30) appearing on the right-hand sides of the respective equations of systems for a special case where G is a bounded domain $G \subset \mathbb{R}^m$, and

$$z : \mathbb{R}^+ \times G \times [1, \infty) \ni (t, x, \lambda) \mapsto z(t, x, \lambda) \in \mathbb{R}^+$$

is a given monotone decreasing function with respect to the variable λ.

If

$$a_n(t, x) := z(t, x, n) \quad \text{for } (t, x) \in D := \mathbb{R}^+ \times G$$

and

$$\mathcal{S} := \int_G \sum_{n=1}^{\infty} a_n(t, x) dx,$$

$$\mathcal{J} := \int_G \int_1^{\infty} z(t, x, \lambda) dx d\lambda \quad \text{for } t \in \mathbb{R}^+, \tag{6.33}$$

then the sequence of functions

$$\{S_N - I_N\} = \{S_N(t, x) - I_N(t, x)\} \tag{6.34}$$

given by

$$S_N = S_N(t, x) := \sum_{n=1}^{N} a_n(t, x),$$

$$I_N = I_N(t, x) := \int_1^N z(t, x, \lambda) d\lambda \quad \text{for } (t, x) \in D \tag{6.35}$$

is always convergent in the pointwise sense, and

$$0 \le \lim_{N \to \infty} [S_N(t, x) - I_N(t, x)] \le a_1(t, x) \quad \text{for } (t, x) \in D. \tag{6.36}$$

In fact, the sequence (6.34) is decreasing because

$$(S_{N-1} - I_{N-1}) - (S_N - I_N) = (I_N - I_{N-1}) - (S_N - S_{N-1})$$
$$= \int_{N-1}^{N} z(t, x, \lambda)d\lambda - a_N > 0,$$

and is bounded, because

$$a_N > 0 \quad \text{and } a_N \le S_N - I_N \le a_1.$$

Hence, the sequence (6.34) converges in a pointwise manner and (6.36) holds.

Assuming that $a_1 = a_1(t, x)$ is a bounded function and $|a_1(t, x)| \le \tilde{a}_1$ in $\overline{D}$, we obtain the following estimate:

$$0 \le \lim_{N \to \infty} [S_N(t, x) - I_N(t, x)] \le \tilde{a}_1 \quad \text{for } (t, x) \in \overline{D}. \tag{6.37}$$

The above-mentioned properties of the sequence $\{S_N(t, x) - I_N(t, x)\}$ may, under appropriate assumptions, extend onto the sequences $\{\mathcal{S}_N(t) - \mathcal{J}_N(t)\}$, where

$$\mathcal{S}_N = \mathcal{S}_N(t) \quad := \quad \int_G \sum_{n=1}^{N} a_n(t, x)dx,$$

$$\mathcal{J}_N = \mathcal{J}_N(t) \quad := \quad \int_G \int_1^{N} z(t, x, \lambda)dxd\lambda \quad \text{for } t \in \mathbb{R}^+, \tag{6.38}$$

and from these, further still, onto the right-hand sides of the equations. This means that the right-hand sides of considered systems may be both identical or different. The latter implies that the two infinite systems of equations may both have solutions, but in general, different ones. This means that solutions of infinite countable and infinite uncountable systems of equations which are the discrete and continuous models of this same process, respectively, may be both identical or different. The above statement is a material result proving that dynamics of the continuous case differ from that of the discrete case.

The next question: Under what assumptions are both models identical and are the assumptions restrictive? Unfortunately, it is difficult to answer these questions. Moreover, it should be emphasized that estimate (6.36) only refers to a term of the reaction function. The difference between continuous versus discrete models also occurs when the mathematical problem is linear.

A very important problem is the comparison between the infinite countable systems and the finite truncated systems of these equations. The comparison between infinite countable systems of equations (precisely: between the solution of initial-boundary value problems) and finite systems of these equations is possible under suitable assumptions. The basic tools of proving this are comparison theorems and maximum principles for finite and infinite systems of equations, the positivity lemma and the Gronwall-Bellman lemma.

6.5 Conclusion

Concluding our considerations, we see that approximations of the solution $z - z(t, x)$ of the infinite system of equations are realized in two steps.

First, we replace the solution $z = z(t, x)$ with the solutions $z_N = z_N(t, x)$ of the finite truncated systems where N is such that

$$\|z - z_N\| \leq \frac{\varepsilon}{2} \quad \text{(truncation error)}.$$

Next, we apply a numerical method (e.g., finite-difference method) to finite systems to compute numerical approximations $z^h_{N,numer} = z^h_{N,numer}(t, x)$ such that

$$\|z_N - z^h_{N,numer}\| \leq \frac{\varepsilon}{2} \quad \text{(numerical method error)}.$$

That is to say, the use of finite differences to approximate infinite systems, as shown using the truncation method, brings about solutions that are reasonably close to each other.

Acknowledgments

We acknowledge the kind permission of World Scientific Publishing Co Pte Ltd to allow us to reproduce in its entirety the following material: Brzychczy, S. and Poznanski, R.R. (2010). Neuronal Models in Infinite-Dimensional Spaces and their Finite-Dimensional Projections, Part 1. *Journal of Integrative Neuroscience*, Volume 9, pp. 11–30.

CHAPTER 7

Fixed Point Method

7.1 Introduction

We consider an infinite system of equations of the form

$$\mathcal{F}^i[z^i](t, x) = f^i(t, x, z(t, x), z), \quad i \in S, \tag{7.1}$$

where

$$\mathcal{F}^i_c := \frac{\partial}{\partial t} - \mathcal{L}^i_c, \quad \mathcal{L}^i_c := \sum_{j,k=1}^m a^i_{jk}(t, x) \frac{\partial^2}{\partial x_j \partial x_k} - \sum_{j=1}^m b^i_j(t, x) \frac{\partial}{\partial x_j} - c^i(t, x)\mathcal{I},$$

$x = (x_1, \ldots, x_m)$, $(t, x) \in (0, T] \times G := D, 0 < T < \infty, G \subset \mathbb{R}^m$, G is an open and bounded domain whose boundary ∂G is an $(m - 1)$-dimensional surface of a class $C^{2+\alpha} \cap C^{2-0}$ for some $\alpha, 0 < \alpha \leq 1$, S is an arbitrary set of indices, $z := z(\cdot, \cdot)$ stands for the mapping

$$z: S \times \overline{D} \to \mathbb{R}, \quad (i, t, x) \mapsto z(i, t, x) = z^i(t, x),$$

composed of unknown functions z^i, and $f^i, i \in S$, are given functions

$$f^i: \overline{D} \times \mathcal{B}(S) \times C_S(\overline{D}) \to \mathbb{R}, \quad (t, x, y, s) \mapsto f^i(t, x, y, s), \quad i \in S,$$

and diagonal operators $\mathcal{F}^i, i \in S$, are uniformly parabolic in $\overline{D} := [0, T] \times \overline{G}$, the right-hand sides f^i of system (7.1) are functionals with respect to the last variable and we do not assume that they are Volterra-type functionals.

For system (7.1) we consider the Fourier first initial-boundary value problem: Find the regular (classical) solution z of system (7.1) in $\overline{D}$ fulfilling the initial-boundary condition

$$z(t, x) = \phi(t, x) \quad \text{for} \quad (t, x) \in \Gamma, \tag{7.2}$$

where Γ is the parabolic boundary of domain D.

Mathematical Neuroscience. http://dx.doi.org/10.1016/B978-0-12-411468-5.00007-7

If $\phi \in C_S^{2+\alpha}(\Gamma)$ and the boundary ∂G is of a class $C^{2+\alpha}$, then without loss of generality we can consider the homogeneous initial-boundary condition

$$z(t, x) = 0 \quad \text{for} \quad (t, x) \in \Gamma \tag{7.3}$$

for system (7.1).

In this chapter, we will study problem (7.1) and (7.3) with use of a topological fixed point method. There are abstract theorems for contraction mappings (Banach theorem) and for compact mappings (Schauder and Leray-Schauder theorems). There are also fixed point theorems for order-preserving operators. This assumption is a well-known condition in the Knaster-Tarski theorem which is based entirely on order.

Before giving three fundamental fixed point theorems, we will define the nonlinear operator.

For a sufficiently regular function $\eta \in C_S(\overline{D})$, we define the nonlinear operator $\mathbf{F} = \{\mathbf{F}^i\}_{i \in S}$ generated by the functions $f^i(t, x, y, s), i \in S$, as follows

$$\mathbf{F}^i : \eta \to \mathbf{F}^i[\eta],$$

where

$$\mathbf{F}^i[\eta](t, x) := f^i(t, x, \eta(t, x), \eta), \quad i \in S. \tag{7.4}$$

We will assume the following properties of the operator $\mathbf{F}$:

Property I   For any $\tau, 0 < \tau \le T$

$$\mathbf{F} : C_S^{0+\alpha}(\overline{D}^\tau) \to C_S^{0+\alpha}(\overline{D}^\tau).$$

Property II   For $u \in C_S^{1+\alpha}(\overline{D}^\tau)$ the following estimate holds

$$\|\mathbf{F}[u]\|_0^{D^\tau} \le B_1 + B_2\|u\|_{1+\alpha}^{D^\tau} \quad \text{for any} \quad \tau, \quad 0 < \tau \le T, \quad \text{(Perron's condition)}, \tag{7.5}$$

where constants $B_1, B_2 > 0$ are independent of τ.

Property III   The operator $\mathbf{F}$ is continuous in the space $C_S^{1+\alpha}(\overline{D})$ in the following sense: if $u_n, u \in C_S^{1+\alpha}(\overline{D}), n \in \mathbb{N}$, and

$$\lim_{n \to \infty} \|u_n - u\|_{1+\alpha} = 0,$$

then

$$\lim_{n \to \infty} \|\mathbf{F}[u_n] - \mathbf{F}[u]\|_0 = 0.$$

Note that if the functions $f^i, i \in S$, generating the nonlinear operator $\mathbf{F}$ satisfy conditions (H_f) and (L) (see Section 2.7), then (see Lemma 4.1)

$$\mathbf{F} : C_S^{0+\alpha}(\overline{D}) \to C_S^{0+\alpha}(\overline{D}).$$

7.2 Theorems on Fixed Point Mapping

Let T be a transformation defined on a set $\mathcal{Y}$ in a Banach space $\mathcal{X}$ and suppose that T maps $\mathcal{Y}$ into itself, i.e., $T[x] \in \mathcal{Y}$ if $x \in \mathcal{Y}$. T is called a *contraction* in $\mathcal{Y}$ if there exists a positive number $\theta < 1$ such that

$$\|T[x] - T[y]\| \leq \theta \|x - y\|$$

for all x, y in $\mathcal{Y}$.

Theorem 7.1 (Banach contraction principle) *Let T be a transformation which maps a closed set $\mathcal{X}_0$ of a Banach space $\mathcal{X}$ into itself, and assume that T is a contraction in $\mathcal{X}_0$. Then there exists the unique point $y \in \mathcal{X}_0$ such that $T[y] = y$ (this point y is called a fixed point of the transformation T).*

A subset $\mathcal{Y}$ of a Banach space $\mathcal{X}$ is said to be *precompact* if for any sequence $\{x_n\}$ in $\mathcal{Y}$ there exists a subsequence $\{x_{n_k}\}$ convergent to some element x of $\mathcal{X}$, not necessarily in $\mathcal{Y}$.

If for any sequence $\{x_n\}$ in $\mathcal{Y}$ there exists a subsequence which converges to an element of $\mathcal{Y}$, then $\mathcal{Y}$ is called a *compact* set.

Theorem 7.2 (Schauder fixed point theorem) *Let $\mathcal{Y}$ be a closed convex subset of a Banach space $\mathcal{X}$ and let T be a continuous transformation on $\mathcal{Y}$ such that $T[\mathcal{Y}]$ is contained in $\mathcal{Y}$ and $T[\mathcal{Y}]$ is precompact. Then there exists a fixed point for T, i.e., a point $y_0 \in \mathcal{Y}$ such that $T[y_0] = y_0$.*

A *transformation* T defined on a subset $\mathcal{Y}$ of $\mathcal{X}$, with $T[\mathcal{Y}] \subset \mathcal{X}$, is said to be *compact* if it maps bounded subsets of $\mathcal{Y}$ into compact subsets of $\mathcal{X}$.

Theorem 7.3 (Leray-Schauder fixed point theorem) *Consider a transformation*

$$T : \mathcal{X} \to \mathcal{Y}, \quad x \mapsto y = T[x, k],$$

where x, y belong to a Banach space $\mathcal{X}$ and k is a real parameter $a \leq k \leq b$.

Assume that

  i. *$T[x, k]$ is defined for all $x \in \mathcal{X}$, $a \leq k \leq b$;*
 ii. *for any fixed k, $T[x, k]$ is continuous in $\mathcal{X}$, i.e., for any $x_0 \in \mathcal{X}$ and for any $\varepsilon > 0$ there exists a $\delta > 0$ such that*

$$\|T[x, k] - T[x_0, k]\| < \varepsilon \quad \text{if} \quad \|x - x_0\| < \delta;$$

iii. *for x in a bounded subset of $\mathcal{X}$, $T[x, k]$ is uniformly continuous in k, i.e., for any bounded set $\mathcal{X}_0 \subset \mathcal{X}$ and for any $\varepsilon > 0$ there exists a δ such that if $x \in \mathcal{X}_0$, $|k_1 - k_2| < \delta$, $k_1, k_2 \in [a, b]$, then $\|T[x, k_1] - T[x, k_2]\| < \varepsilon$;*
 iv. *for any fixed k, $T[x, k]$ is a compact transformation, i.e., it maps bounded subsets of $\mathcal{X}$ into compact subsets of $\mathcal{X}$;*

> **v.** *there exists a constant M such that every possible solution x of x − T[x, k] = 0,*
> *x ∈ X, a ≤ k ≤ b satisfies:* $\|x\| \le M$;
>
> **vi.** *the equation x − T[x, a] = 0 has the unique solution in X.*

Under these assumptions, there exists a solution of the equation x − T[x, b] = 0.

7.3 Banach Theorem for Contraction Mappings

To prove the existence and uniqueness of the solution of problem (7.1) and (7.3), we shall apply the Banach fixed point theorem for contraction mappings. Considering mainly Banach spaces of continuous mappings, we give some natural sufficient conditions for existence and uniqueness. Unfortunately the existence of solutions is assured exclusively in a part D^τ of the given domain D contained in the zone $\{(t, x) : 0 < t < \tau, \; x \in \mathbb{R}^m\}$, where $\tau, 0 < \tau \le T$, is a sufficiently small number.

Theorem 7.4 *Let assumptions (H_f), (H_a), (L) (see Section 2.7) hold and τ^* be a sufficiently small number. Then there exists the unique regular solution z of the initial-boundary value problem (7.1) and (7.3) in domain $\overline{D}^\tau$, where $0 < \tau < \tau^* \le T$, in the space $C_S^{1+\beta}(\overline{D}^\tau)$. Moreover, $z \in C_S^{2+\alpha}(\overline{D}^\tau)$.*

Proof. Denote by

$$\mathbf{A_1} := \{w : w \in C_S^{1+\alpha}(\overline{D}^\tau), \quad w(t, x) = 0 \quad \text{on} \quad \Gamma^\tau, \quad 0 < \tau \le T, \quad 0 < \alpha < 1\}.$$

$\mathbf{A_1}$ is the closed subset of the space $C_S^{1+\alpha}(\overline{D}^\tau)$.

For $u \in \mathbf{A_1}$ we define a transformation $\mathbf{T_1}$ setting

$$z = \mathbf{T_1}[u],$$

where z is the (supposedly unique) solution of the infinite systems of linear equations of the form

$$\mathcal{F}^i[z^i](t, x) = f^i(t, x, u(t, x), u(\cdot, \cdot)) := \mathbf{F}^i[u](t, x) \quad i \in S \quad \text{in} \quad D^\tau, \tag{7.6}$$

with the homogeneous boundary condition

$$z(t, x) = 0 \quad \text{on} \quad \Gamma^\tau. \tag{7.7}$$

Corollary 7.1 Observe that system (7.6) and (7.7) has the following crucial property: for an arbitrary $i \in S$ the ith equation of this infinite system of equations depends on the ith unknown function $z^i = z^i(t, x)$ only. Therefore, this system is a collection of several equations each of which is an equation in one unknown function only.

Two situations are possible. If system (7.1) is infinite countable, then $\overline{\overline{S}} = \overline{\overline{\mathbb{N}}} = \aleph_0$.[1] In this system an arbitrary fixed index $j \in \mathbb{N}$, the jth equation of system (7.6) and (7.7) depends on the one unknown function z^j where $z = \{z^j\}_{j\in\mathbb{N}}$ and $z^j = z^j(t, x) := z(t, x, j)$ only. In this case we consider the infinite countable system of equations of the form:

$$\mathcal{F}^j[z^j](t, x) = f^j(t, x, u(t, x), u(\cdot, \cdot)) \quad \text{for} \quad j \in \mathbb{N} \quad \text{in} \quad D, \tag{7.8}$$

which is a collection of the infinite countable number of separate scalar problems.

These infinite countable systems appear in discrete models of processes considered.

Analogously, if system (7.1) is an infinite uncountable system, then $\overline{\overline{S}} = \overline{\overline{\mathbb{R}}} = \overline{\overline{\Lambda}} = \mathfrak{c}$ where $\Lambda \subset \mathbb{R}$ is the compact set of indices (with an arbitrary number of elements, e.g., $\Lambda := [\lambda_0, \lambda_1] \subset \mathbb{R}$). In this system for an arbitrary fixed parameter $\lambda \in \Lambda$, the λth equation of this system depends on the one unknown function z^λ only, where $z = \{z^\lambda\}_{\lambda\in\Lambda}$ and $z^\lambda = z^\lambda(t, x) = z(t, x, \lambda)$.

In this case we consider the infinite uncountable system of equations of the form

$$\mathcal{F}^\lambda[z^\lambda](t, x) := f^\lambda(t, x, u(t, x, \lambda), u(\cdot, \cdot, \lambda)), \quad \text{for} \quad \lambda \in \Lambda \quad \text{in} \quad D, \tag{7.9}$$

which is a collection of separate equations of the infinite uncountable number of scalar problems. These infinite uncountable systems are continuous models of described processes.

For $u \in \mathbf{A_1}$, by Lemma 4.1, there is $\mathbf{F}^i[u] \in C_S^{0+\alpha}(\overline{D}^\tau)$ and by Lemma 4.2, system (7.6) and (7.7) has the unique regular solution z and $z \in C_S^{2+\alpha}(\overline{D}^\tau)$. Moreover, by Theorem A.2, $z \in C_S^{1+\beta}(\overline{D}^\tau)$ for arbitrary $\beta, 0 < \beta < 1$. Therefore, the transformation $\mathbf{T_1}$ maps the set $\mathbf{A_1}$ into itself.

Let us choose an arbitrary exponent $\beta, 0 < \alpha < \beta < 1$. We prove now that $\mathbf{T_1}$ is a contraction in $C_S^{1+\beta}(\overline{D}^\tau)$.

Let $u, \tilde{u} \in \mathbf{A_1}$ and $z = \mathbf{T_1}[u], \tilde{z} = \mathbf{T_1}[\tilde{u}]$. According to the definition of $\mathbf{T_1}$, there is

$$\begin{cases} \mathcal{F}^i[z^i - \tilde{z}^i](t, x) = \mathbf{F}^i[u](t, x) - \mathbf{F}^i[\tilde{u}](t, x), & i \in S \quad \text{in} \quad D^\tau, \\ z(t, x) - \tilde{z}(t, x) = 0 \quad \text{on} \quad \Gamma^\tau. \end{cases} \tag{7.10}$$

---

[1] We use $\overline{\overline{S}}$ to denote the cardinality of the set S. The symbol $\aleph_0$ denotes the cardinality of the set of nonnegative integers; that is, $\overline{\overline{\mathbb{N}}} = \aleph_0$. The symbol $\aleph$ (aleph) is the first letter of the Hebrew alphabet. The symbol $\aleph_0$ is pronounced "aleph zero." An infinite set of cardinality $\aleph_0$ is called infinite countable. The cardinality of the set of reals is denoted by $\mathfrak{c}$; that is, $\overline{\overline{\mathbb{R}}} = \mathfrak{c}$. A set of cardinality $\mathfrak{c}$ is called a set of continuum cardinality. It is in particular an infinite uncountable set.

Because $z - \tilde{z}$ is the solution of problem (7.10), we can apply Theorem A.2. From estimate A.2.5 and condition (L) we obtain

$$\|z - \tilde{z}\|_{1+\beta}^{D^\tau} \leq \overline{K}\tau^{\frac{1-\beta}{2}} \left\|\mathbf{F}^i[u] - \mathbf{F}^i[\tilde{u}]\right\|_0^{D^\tau}$$

$$= \overline{K}\tau^{\frac{1-\beta}{2}} \sup_{\substack{(t,x)\in D^\tau \\ i\in S}} |f^i(t, x, u(t, x), u) - f^i(t, x, \tilde{u}(t, x), \tilde{u})|$$

$$\leq \overline{K}\tau^{\frac{1-\beta}{2}} \left(L_1 \|u(t, x) - \tilde{u}(t, x)\|_{B(S)}^{D^\tau} + L_2 \|u - \tilde{u}\|_0^{D^\tau}\right)$$

$$\leq \overline{K}(L_1 + L_2)\tau^{\frac{1-\beta}{2}} \|u - \tilde{u}\|_0^{D^\tau} \leq \overline{K}(L_1 + L_2)\tau^{\frac{1-\beta}{2}} \|u - \tilde{u}\|_{1+\beta}^{D^\tau} \text{ for } 0 < \tau \leq T.$$

If we now assume

$$\theta := \overline{K}(L_1 + L_2)\tau^{\frac{1-\beta}{2}} < 1,$$

or

$$0 < \tau < \min\left\{(\overline{K}(L_1 + L_2))^{\frac{2}{\beta-1}}, T\right\} := \tau^*, \tag{7.11}$$

then we get

$$\|z - \tilde{z}\|_{1+\beta}^{D^\tau} \leq \theta\|u - \tilde{u}\|_{1+\beta}^{D^\tau}.$$

This means that the transformation $\mathbf{T}_1$ is a contraction in $C_S^{1+\beta}(\overline{D}^\tau)$ for $0 < \tau < \tau^*$, where τ^* is a sufficiently a small number defined by (7.11).

Therefore, by the Banach contraction principle, the transformation $\mathbf{T}_1$ has the unique fixed point $z \in \mathbf{A}_1$. If we assume on the contrary that $\mathbf{T}_1[\tilde{z}] = \tilde{z}$ and $\mathbf{T}_1[z^*] = z^*$ and $\tilde{z} \neq qz^*$, then we get

$$\|\tilde{z} - z^*\|_{1+\alpha} = \|\mathbf{T}_1[\tilde{z}] - \mathbf{T}_1[z^*]\|_{1+\alpha} \leq \theta\|\tilde{z} - z^*\|_{1+\alpha}.$$

Because $0 < \theta < 1$, we conclude that $\|\tilde{z} - z^*\|_{1+\alpha} = 0$, which is a contradiction. It is obvious that z is the weak C-solution of problem (7.1), and (7.3) in $\overline{D}^\tau$.

Let $\tilde{z} \in C_S^{1+\alpha}(\overline{D}^\tau)$ be the unique C-solution of problem (7.1) and (7.3). We consider the following problem

$$\begin{cases} \mathcal{F}^i[z^i](t, x) = \mathbf{F}^i[\tilde{z}](t, x), & i \in S, \text{ in } D^\tau, \\ z(t, x) = 0 \text{ on } \Gamma^\tau. \end{cases} \tag{7.12}$$

Applying Lemma 4.1, we have $\mathbf{F}^i[\tilde{z}] \in C^{0+\alpha}(\overline{D}^\tau)$, and by Lemma 4.2, problem (7.12) has the unique regular solution $z \in C_S^{2+\alpha}(\overline{D}^\tau)$ in domain $\overline{D}^\tau$, which ends the proof. $\square$

Corollary 7.2 The unique solution of problem (7.1) and (7.3) can be expressed with the use of the method of successive approximations. Let the successive terms of the approximations sequence $\{z_n\}$ be defined as

$$z_1 = \mathbf{T}_1[z_0], \quad z_{n+1} = \mathbf{T}_1[z_n], \quad n = 1, 2, \ldots,$$

where z_0 is an arbitrary point $z_0 \in C_S^{1+\beta}(\overline{D})$ (e.g., $z_0 \in C_S^{2+\alpha}(\overline{D})$ is an extension of ϕ from Γ onto $\overline{D}$).

Since $\mathbf{T_1}$ is a contraction, we can conclude that $\{z_n\}$ satisfies the Cauchy condition. Therefore, from the completeness of the space $C_S^{1+\beta}(\overline{D})$, we deduce that there exists the unique point $\tilde{z} \in C_S^{1+\beta}(\overline{D})$ such that:

$$\lim_{n \to \infty} z_n^i(t, x) = \tilde{z}^i(t, x), \quad i \in S.$$

Now we give the existence and uniqueness theorem on a weak C-solution of homogeneous initial-boundary value problem for an infinite system of equations of the form

$$\begin{cases} \mathcal{T}^i[z^i](t, x) - \tilde{f}^i(t, x, z), & i \in S, \quad \text{in } D, \\ z(t, x) = 0 \quad \text{on } \Gamma \end{cases} \tag{7.13}$$

obtained by means of the Banach contraction principle with the use of *Bielecki's weighted norm*[2] in whole domain $\overline{D}$.

Theorem 7.5 *Let assumptions* (H_a), (H_f), *and* (L) *(see Section 2.7) hold. Then there exists the unique global weak C-solution z of problem* (7.13) *in* $\overline{D}$.

Proof. Denote

$$\mathbf{A_0} := \{w : w \in C_S(\overline{D}), w(t, x) = 0 \quad \text{on } \Gamma\}.$$

For $u \in \mathbf{A_0}$ we define a transformation $\mathbf{T_0} = \{\mathbf{T}_0^i\}_{i \in S}$

$$u \mapsto \mathbf{T}_0^i[u] = z^i, \quad i \in S,$$

where z is the (supposedly unique) solution of the following linear homogeneous problem

$$\begin{cases} \mathcal{F}^i[z^i](t, x) = \tilde{f}^i(t, x, u), & i \in S, \quad \text{in } D, \\ z(t, x) = 0 \quad \text{on } \Gamma. \end{cases}$$

---

[2] Let a function $\psi : [0, T] \to \mathbb{R}_+$, $\psi \in C([0, T])$ satisfy the inequality

$$\psi(t) \geq \frac{L}{\theta} \int_0^t \psi(\tau) d\tau,$$

with some $\theta \in (0, 1)$ and constant $L > 0$.
We define a new norm $\|\cdot\|_{0,\psi}$ in the space $C_S(\overline{D})$ in the following way

$$\|u\|_{0,\psi} := \left\| \frac{u}{\psi} \right\|_0 \quad \text{for } u \in C_S(\overline{D}),$$

where $\frac{u}{\psi}$ is given by $\left(\frac{u}{\psi}\right)(t, x) = \frac{u(t,x)}{\psi(t)}$ and $\frac{u}{\psi} = \left\{\frac{u^i}{\psi}\right\}_{i \in S}$.
It is clear that the supremum norm $\|\cdot\|_0$ in $C_S(\overline{D})$ is equivalent to the weighted norm $\|\cdot\|_{0,\psi}$.

By assumption (H_a), (H_f) (see Section 2.7) we will write this transformation by means of the integral formula

$$z^i(t, x) := \mathbf{T_0}^i[u](t, x) = \int_0^t \int_G \mathcal{G}^i(t, x; \tau, \xi) \tilde{f}^i(\tau, \xi, u) d\tau d\xi, \quad i \in S,$$

where $\mathcal{G}^i(t, x; \tau, \xi)$, $i \in S$, are Green's functions for these problems in domain D. It is clear that $\mathbf{T_0} \colon \mathbf{A_0} \to \mathbf{A_0}$.

If $u, \tilde{u} \in \mathbf{A_0}$ then by assumption (L) and properties of Green's function, there is

$$\left| \mathbf{T}_0^i[u](t, x) - \mathbf{T}_0^i[\tilde{u}](t, x) \right| \leq \int_0^t \int_G \mathcal{G}^i(t, x; \tau, \xi) \left| \tilde{f}^i(\tau, \xi, u) - \tilde{f}^i(\tau, \xi, \tilde{u}) \right| d\tau d\xi$$

$$\leq \int_0^t \int_G \mathcal{G}^i(t, x; \tau, \xi) L \psi(\tau) \left\| \frac{u - \tilde{u}}{\psi} \right\|_0 d\tau d\xi$$

$$\leq \int_0^t L \psi(\tau) \|u - \tilde{u}\|_{0,\psi} d\tau \leq \theta \psi(t) \|u - \tilde{u}\|_{0,\psi},$$

where $0 < \theta < 1, i \in S$.

Finally

$$\left\| \mathbf{T}_0[u] - \mathbf{T}_0[\tilde{u}] \right\|_{0,\psi} \leq \theta \|u - \tilde{u}\|_{0,\psi}$$

for $u, \tilde{u} \in C_S(\overline{D})$.

According to the Banach contraction principle, there exists the unique fixed point $z \in \mathbf{A_0}$ of the transformation $\mathbf{T_0}$. Therefore, the function z is the weak C-solution of problem (7.13) in whole $\overline{D}$. This completes the proof. $\qquad \square$

7.4 Schauder Fixed Point Theorem for Compact Mappings

To prove the existence of solutions of problem (7.1) and (7.3) in $\overline{D}$, we shall apply the Schauder fixed point theorem. Considering mainly Banach spaces of continuous and bounded functions, we give some natural sufficient conditions for the existence of solutions.

Theorem 7.6 *Let assumptions (H_f), (H_a), (H_ϕ) hold (see Section 2.7), the operator $\mathbf{F}[u]$ have Properties I–III, and $\tau^* \leq T$, be a sufficiently small number. Then there exists the solution z of problem (7.1) and (7.3) in domain $\overline{D}^\tau$, where for an arbitrary fixed $\tau, 0 < \tau < \tau^* \leq T$, we define*

$$D^\tau := \{(t, x) : 0 < t \leq \tau, \ x \in G\},$$

in the space $C_S^{1+\beta}(\overline{D}^\tau)$, and $z \in C_S^{2+\alpha}(\overline{D}^\tau)$.

Proof.   Denote by

$$\mathbf{A_2} := \left\{ u \colon u \in C_S^{1+\alpha}(\overline{D}^\tau), \|u\|_{1+\alpha}^{D^\tau} \le M, \quad u(t, x) = 0 \quad \text{on} \quad \Gamma^\tau, \quad 0 < \tau \le T \right\},$$

the closed ball in the space $C_S^{1+\alpha}(\overline{D}^\tau)$ where $M > 0$ is some constant and $0 < \alpha < 1$.
For $u \in \mathbf{A_2}$, we define a transformation $\mathbf{T_2}$ setting

$$z = \mathbf{T_2}[u],$$

where z is a regular solution of the linear problem

$$\begin{cases} \mathcal{F}^i[z^i](t, x) = \mathbf{F}^i[u](t, x), \quad i \in S \quad \text{in} \quad D^\tau, \\ z(t, x) = 0 \quad \text{on} \quad \Gamma^\tau. \end{cases} \tag{7.14}$$

From Property I and Theorem A.1 it follows that for $u \in \mathbf{A_2}$ problem (7.14) has the unique solution $z \in C_S^{2+\alpha}(\overline{D}^\tau)$. Therefore, $z \in C_S^{1+\beta}(\overline{D}^\tau)$ and transformation $\mathbf{T_2}$ is well defined.

Now we prove that $\mathbf{T_2}$ maps $\mathbf{A_2}$ into itself. By Theorem A.2, for any positive θ, $0 < \theta < 1$, there exists a constant $\overline{K} = \overline{K}(\theta)$ such that estimate A.2.5 holds, i.e.,

$$\|z\|_{1+\theta}^{D^\tau} \le \overline{K} \tau^{\frac{1-\theta}{2}} \|\mathbf{F}[u]\|_0^{D^\tau} \tag{7.15}$$

for $0 < \tau \le T$.

Hence by Property I

$$\|z\|_{1+\theta}^{D^\tau} \le \overline{K} \tau^{\frac{1-\theta}{2}} B \|u\|_1^{D^\tau} \le \overline{K} \tau^{\frac{1-\theta}{2}} B \|u\|_{1+\alpha}^{D^\tau}.$$

If we assume

$$\overline{K} \tau^{\frac{1-\theta}{2}} B \le 1,$$

or

$$0 < \tau \le \tau^* := \min \left\{ (\overline{K} B)^{\frac{2}{\theta-1}}, T \right\} \tag{7.16}$$

then we get

$$\|z\|_{1+\theta}^{D^\tau} \le M.$$

Therefore, the transformation $\mathbf{T_2}$ maps the set $\mathbf{A_2}$ into itself, i.e.,
$\mathbf{T_2}(\mathbf{A_2}) := \{\mathbf{T_2}[u] \colon u \in \mathbf{A_2}\} \subset \mathbf{A_2}$ for $\tau, 0 < \tau \le \tau^*$.

Now let $\beta = \theta$ and $0 < \alpha < \beta < 1$. The set $\mathbf{T_2}(\mathbf{A_2})$ is a bounded subset of the space $C_S^{1+\beta}(\overline{D}^\tau)$; therefore, this set is a precompact subset of $C_S^{1+\alpha}(\overline{D}^\tau)$.

To prove that $\mathbf{T_2}$ is continuous, we note that if $u_\nu, u \in \mathbf{A_2}$ and $z_\nu = \mathbf{T_2}[u_\nu]$, $z = \mathbf{T_2}[u]$ then, from the definition of $\mathbf{T_2}$, we have

$$\begin{cases} \mathcal{F}^i[z_\nu^i - z^i](t, x) = \mathbf{F}^i[u_\nu](t, x) - \mathbf{F}^i[u](t, x), & i \in S \text{ in } D^\tau, \\ z_\nu(t, x) - z(t, x) = 0 & \text{on } \Gamma^\tau. \end{cases}$$

Applying the estimate (7.15) to this problem, we obtain

$$\|\mathbf{T_2}[u_\nu] - \mathbf{T_2}[u]\|_{1+\beta}^{D^\tau} = \|z_\nu - z\|_{1+\beta}^{D^\tau} \leq \overline{K}\tau^{\frac{1-\beta}{2}} \|\mathbf{F}[u_\nu] - \mathbf{F}[u]\|_0^{D^\tau}.$$

If now

$$\lim_{\nu \to \infty} \|u_\nu - u\|_{1+\alpha}^{D^\tau} = 0,$$

then by Property II there is

$$\lim_{\nu \to \infty} \|\mathbf{F}[u_\nu] - \mathbf{F}[u]\|_0^{D^\tau} = 0.$$

Therefore by Property III

$$\lim_{\nu \to \infty} \|\mathbf{T_2}[u_\nu] - \mathbf{T_2}[u]\|_{1+\beta}^{D^\tau} = 0,$$

i.e., the transformation $\mathbf{T_2}$ is continuous.

Using the Schauder fixed point theorem we conclude that the transformation $\mathbf{T_2}$ has a fixed point $z \in \mathbf{A_2}$. z is then a solution of problem (7.1) and (7.2) in the space $C_S^{1+\beta}(\overline{D^\tau})$ for $0 < \tau \leq \tau^*$, where τ^* defined by (7.16) is a sufficiently small number. From Theorem A.1 it follows that z also belongs to $C_S^{2+\alpha}(\overline{D^\tau})$. This completes the proof. □

Remark 7.1 If we additionally suppose that there exists a positive constant M_0 such that, for any $M > M_0$, we have

$$K\|\mathbf{F}[u]\|_0 \leq M \quad \text{in} \quad \overline{D}$$

for all functions $u \in C_S^{1+\alpha}(\overline{D})$ satisfying $\|u\|_{1+\alpha}^D \leq M$ where K is the constant appearing in Theorem A.2, then problem (7.1) and (7.3) has a solution in the whole domain $\overline{D}$.

7.5 Leray-Schauder Theorem for Compact Mappings

To prove the existence of global solutions, we apply the Leray-Schauder fixed point theorem. First we extend some *a priori* estimates of the Friedman type, and next we prove the existence theorem for the whole domain $\overline{D}$.

Theorem 7.7 *Let assumptions (H_a), (H_f), and (H_ϕ) hold (see Section 2.7) and the function z be a unique solution of problem (7.1) and (7.2) in $\overline{D}$. Then, for any $\beta, 0 < \beta < 1, z \in C_S^{1+\beta}(\overline{D})$, and*

$$\|z\|_{1+\beta} \leq K, \tag{7.17}$$

where $K > 0$ is a constant depending on the same parameter that $\overline{K}$ does.

Proof. Let z be a solution of problem (7.1) and (7.2) in $\overline{D}$ and the function $\Phi \in C_S^{1+\beta}(\overline{D}) \cap C_S^{1,2}(\overline{D})$ be such an extension of ϕ onto $\overline{D}$ that

$$\|\Phi\|_{1+\beta} \leq A_3, \quad \|\Phi\|_{1,2} \leq A_4. \tag{7.18}$$

Then the function

$$v(t, x) = z(t, x) - \Phi(t, x)$$

is the solution of the following homogeneous problem

$$\begin{cases} \mathcal{F}^i[v^i](t, x) = \mathbf{F}^i[z](t, x) - \mathcal{F}^i[\Phi^i](t, x) := g^i(t, x), \quad i \in S \quad \text{in} \quad D, \\ v(t, x) = 0 \quad \text{on} \quad \Gamma. \end{cases} \tag{7.19}$$

The functions $g = \{g^i : i \in S\}$ are continuous in $\overline{D}$ and vanish on ∂G. Therefore, by Theorem A.2, for any $\beta, 0 < \beta < 1$, there exists a constant $\overline{K} > 0$, depending only on $\beta, \mu_0, A_1, A_2, A_3, A_4, B_1, B_2$ and on the domain D (but not on τ) such that the following estimate holds

$$\|v\|_{1+\beta}^{D^\tau} \leq \overline{K} \tau^{\frac{1-\beta}{2}} \|g\|_0^{D^\tau} = \overline{K} \tau^{\frac{1-\beta}{2}} \|\mathbf{F}^i[z] - \mathcal{F}^i[\Phi^i]\|_0^{D^\tau}$$

$$\leq \overline{K} \tau^{\frac{1-\beta}{2}} \left(\|\mathbf{F}^i[v + \Phi]\|_0^{D^\tau} + \|\mathcal{F}^i[\phi^i]\|_0^{D^\tau} \right) \tag{7.20}$$

for $0 < \tau \leq T$. From (7.18) applied to (7.20), there follows

$$\|v\|_{1+\beta}^{D^\tau} \leq \overline{K} \tau^{\frac{1-\beta}{2}} \left(B_1 + B_2 \|v + \Phi\|_{1+\beta}^{D^\tau} \right) + \|\mathcal{F}^i[\Phi^i]\|_0^{D^\tau} \right)$$

$$\leq \overline{K} \tau^{\frac{1-\beta}{2}} \left(B_1 + B_2 \left(\|v\|_{1+\beta}^{D^\tau} + \|\Phi\|_{1+\beta}^{D^\tau} \right) + \left\| \mathcal{F}^i[\Phi^i] \right\|_0^{D^\tau} \right) \tag{7.21}$$

or

$$\left(1 - \overline{K} B_2 \tau^{\frac{1-\beta}{2}} \right) \|v\|_{1+\beta}^{D^\tau} \leq \overline{K} \tau^{\frac{1-\beta}{2}} \left(B_1 + B_2 \|\Phi\|_{1+\beta}^{D^\tau} + \left\| \mathcal{F}^i[\Phi^i] \right\|_0^{D^\tau} \right).$$

If now

$$1 - \overline{K} B_2 T^{\frac{1-\beta}{2}} > 0$$

then from (7.21) we get the estimate

$$\|v\|_{1+\beta} \leq K \tag{7.22}$$

in the whole domain D, where $K > 0$ is a suitable constant, which finishes the proof.

If nevertheless

$$1 - \overline{K} B_2 T^{\frac{1-\beta}{2}} \leq 0$$

then estimate (7.22) holds only in the zone D^τ of the domain D, i.e.,

$$\|v\|_{1+\beta}^{D^\tau} \leq K, \tag{7.23}$$

where $0 < \tau \leq \tau^* < (\overline{K} B_2)^{\frac{2}{\beta - 1}}$.

In this case we proceed step by step to estimate (7.22) in the whole domain D.

Let $\sigma, 0 < \sigma \leq \tau^*$, be such a number that T is an integer multiple of σ, i.e., $T = \kappa\sigma$ where $\kappa \in \mathbb{N}$.

We remark that if one considers homogeneous problem (7.19) in a suitable domain, i.e.,

$$
\begin{cases}
\mathcal{F}^i[w^i](t, x) = g^i(t, x), & i \in S, \quad \text{in } D^{(\sigma)} := D^{\rho+\sigma} - D^{\rho}, \\
w(t, x) = 0 & \text{on } \Gamma^{(\sigma)} := \Gamma^{\rho+\sigma} \setminus \Gamma^{\rho} \cup D \cap \{t = \rho\},
\end{cases}
\tag{7.24}
$$

for any $\rho, 0 \leq \rho \leq T - \sigma$, then from Theorem A.2 it follows that

$$
\|w\|_{1+\beta}^{D^{(\sigma)}} \leq K_\rho \sigma^{\frac{1-\beta}{2}} \|g\|_0^{D^{(\sigma)}} = \overline{K}_\rho \|g\|_0^{D^{(\sigma)}}.
\tag{7.25}
$$

Let $\xi(t)$ be a continuously differentiable function in t, vanishing for $t \leq \frac{1}{2}$ and equal to 1 for $t \geq 1$.

Consider the functions

$$
\tilde{v}_\nu^i(t, x) = \xi\left(\frac{t}{\sigma} - \frac{\nu - 1}{2}\right) v^i(t, x), \quad i \in S, \quad \text{for } 0 \leq \nu \leq 2(\kappa - 1),
$$

in the domains $D_\nu^{(\sigma)} := D \cap \left\{\nu\frac{\sigma}{2} < t < (\nu + 2)\frac{\sigma}{2}\right\}$ where v is the solution of problem (7.19).

These functions satisfy the system of equations

$$
\begin{aligned}
\mathcal{F}^i[\tilde{v}_\nu^i](t, x) &= \frac{d\xi}{dt} v^i(t, x) + \xi \frac{\partial v^i}{\partial t} - \xi \mathcal{L}^i[v^i](t, x) \\
&= \frac{d\xi}{dt} v^i(t, x) + \xi \mathcal{F}^i[v^i](t, x) = \frac{d\xi}{dt} v^i(t, x) + \xi g^i(t, x) \\
&:= g_\nu^i(t, x), \quad i \in S \quad \text{in } D_\nu^{(\sigma)}
\end{aligned}
$$

with the initial-boundary condition

$$
\tilde{v}_\nu(t, x) = 0 \quad \text{on } \Gamma_\nu^{(\sigma)} := \Gamma \cap \left\{\nu\frac{\sigma}{2} < t < (\nu + 2)\frac{\sigma}{2}\right\} \cup D \cap \left\{t = \nu\frac{\sigma}{2}\right\}.
$$

Using (7.22) we get the following estimate

$$
\|v\|_{1+\beta}^{D_\nu^{(\sigma)}} \leq \overline{K}_\nu \|g_\nu\|_0^{D_\nu^{(\sigma)}} \leq K_\nu \quad \text{for } 0 \leq \nu \leq 2(\kappa - 1) \quad \text{in } D_\nu^{(\sigma)}.
$$

Therefore, by using (7.24) we obtain

$$
\|v\|_{1+\beta} \leq K \quad \text{in the whole domain } D,
$$

where $K = \max\{K_0, K_1, \ldots, K_{2(\kappa-1)}\}$. This completes the proof. $\qquad\square$

Theorem 7.8 *Let assumptions* (H_a), (H_f), (H_ϕ) *(see Section 2.7) and Properties I and II hold. Then there exists a solution z of problem (7.1) and (7.2) in the domain $\overline{D}$ in the space $C_S^{1+\beta}(\overline{D})$, and $z \in C_S^{2+\alpha}(\overline{D})$.*

Proof. To prove the theorem, we apply the Leray-Schauder fixed point theorem.

We will denote by $\mathbf{A_3}$ the set of functions

$$\mathbf{A_3} := \left\{ u : u \in C_S^{1+\alpha}(\overline{D}), u(t, x) = \phi(t, x) \quad \text{on} \quad \Gamma \right\}.$$

For $u \in \mathbf{A_3}$ and $\mu \in [0, 1]$ we define a transformation $\mathbf{T_3}$ setting

$$z = \mathbf{T_3}[u, \lambda],$$

where z is the (supposedly unique) solution of the following linear initial-boundary value problem

$$\begin{cases} \mathcal{F}^i[z^i](t, x) = \mu \left(\mathbf{F}^i[u](t, x) - \mathcal{F}^i[\Phi^i](t, x) \right) + \mathcal{F}^i[\Phi^i](t, x), \quad i \in S, \quad \text{in} \quad D, \\ z(t, x) = \phi(t, x) \quad \text{on} \quad \Gamma, \end{cases} \quad (7.26)$$

where $\Phi \in C_S^{1+\beta}(\overline{D}) \cap C_S^{2+\alpha}(\overline{D})$ is such an extension of ϕ from Γ onto $\overline{D}$ that (7.18) holds.

i. For $u \in \mathbf{A_3}$, by Property I, we have $\mathbf{F}[u] \in C_S^{0+\alpha}(\overline{D})$. From Theorem A.1 it follows that problem (7.26) has the unique solution z and $z \in C_S^{2+\alpha}(\overline{D})$. Moreover, by Theorem A.2, $z \in C_S^{1+\beta}(\overline{D})$ for any $\beta, 0 < \beta < 1$.
   Therefore, the mapping $\mathbf{T_3}[u, \mu]$ is well defined for all $u \in \mathbf{A_3}$, $\mu \in [0, 1]$.
ii. For any fixed $\mu, \mu \in [0, 1]$, the mapping $\mathbf{T_3}[u, \mu]$ is continuous with respect to u in $\mathbf{A_3}$, i.e., the condition

$$\lim_{n \to \infty} \|u_n - u\|_{1+\alpha} = 0 \quad \text{implies} \quad \lim_{n \to \infty} \|\mathbf{T_3}[u_n, \mu] - \mathbf{T_3}[u, \mu]\|_{1+\alpha} = 0.$$

Indeed, by the definition of the transformation $\mathbf{T_3}$, we have $\mathbf{T_3}[u_n, \mu] = z_n$, where z_n, for $n = 1, 2, \ldots$, are the solutions of the problems

$$\begin{cases} \mathcal{F}^i[z_n^i](t, x) = \mu \left(\mathbf{F}^i[u_n](t, x) - \mathcal{F}^i[\Phi^i](t, x) \right) + \mathcal{F}^i[\Phi^i](t, x), \quad i \in S \quad \text{in} \quad D, \\ z_n(t, x) = \phi(t, x) \quad \text{on} \quad \Gamma. \end{cases}$$
$$(7.27)$$

From (7.26) and (7.27) it follows that

$$\begin{cases} \mathcal{F}^i[z_n^i - z^i](t, x) = \mu \left(\mathbf{F}^i[u_n](t, x) - \mathbf{F}^i[u](t, x) \right), \quad i \in S \quad \text{in} \quad D, \\ z_n(t, x) - z(t, x) = 0 \quad \text{on} \quad \Gamma. \end{cases} \quad (7.28)$$

By Property II we have

$$\lim_{n \to \infty} \|\mathbf{F}[u_n] - \mathbf{F}[u]\|_0 = 0$$

and using Theorem A.2 to problem (7.28) we obtain

$$\lim_{z_n \to \infty} \|z_n - z\|_{1+\alpha} = 0.$$

iii. Let $z_1 = \mathbf{T_3}[u, \mu_1]$ and $z_2 = \mathbf{T_3}[u, \mu_2]$, where $0 < \mu_1, \mu_2 < 1$. Then

$$\begin{cases} \mathcal{F}^i[z_1^i - z_2^i](t, x) = (\mu_1 - \mu_2)\left(\mathbf{F}^i[u](t, x) - \mathcal{F}^i[\Phi^i](t, x)\right), & i \in S, \quad \text{in } D, \\ z_1(t, x) - z_2(t, x) = 0 \quad \text{on } \Gamma. \end{cases}$$

(7.29)

If $\|u\|_{1+\alpha} \le M$ then, from Theorem A.2 applied to problem (7.29), we obtain

$$\begin{aligned} \|\mathbf{T_3}[u, \mu_1] - \mathbf{T_3}[u, \mu_2]\|_{1+\alpha} &= \|z_1 - z_2\|_{1+\alpha} \\ &\le K \left\| (\mu_1 - \mu_2)\left(\mathbf{F}^i[u] - \mathcal{F}^i[\Phi^i]\right) \right\|_0 \\ &\le K |\mu_1 - \mu_2| \left(B_1 + B_2 \|u\|_{1+\alpha} + \left\| \mathcal{F}^i[\Phi^i] \right\|_0 \right) \\ &\le K \left(B_1 + B_2 M + \tilde{A}_4 \right) |\mu_1 - \mu_2| = \tilde{K} |\mu_1 - \mu_2|, \end{aligned}$$

where $\tilde{A}_4 := \left\| \mathcal{F}^i[\Phi^i] \right\|_0$ and $\tilde{K} := K(B_1 + B_2 M + \tilde{A}_4)$.
Therefore, for u in a bounded set of $\mathbf{A_3}$, the mapping $\mathbf{T_3}[u, \mu]$ is uniformly continuous with respect to μ.

iv. Let $0 < \alpha < \beta < 1$. If $\|u\|_{1+\alpha} \le M$ then, by the same argument as that applied above to problem (7.26), we obtain

$$\begin{aligned} \|\mathbf{T_3}[u, \mu]\|_{1+\beta} = \|z\|_{1+\beta} &\le K \|\mu \mathbf{F}^i[u] + (1 - \mu)\mathcal{F}^i[\Phi^i]\|_0 \\ &\le K \left(\mu \|\mathbf{F}^i[u]\|_0 + (1 - \mu)\|\mathcal{F}^i[\Phi^i]\|_0 \right) \\ &\le K \left(\mu(B_1 + B_2 \|u\|_{1+\alpha}) + (1 - \mu)\tilde{A}_4 \right) \\ &\le K \left(\mu(B_1 + B_2 M) + (1 - \mu)\tilde{A}_4 \right) := M^*. \end{aligned}$$

Therefore, $\mathbf{T_3}$ (for any fixed μ, $\mu \in [0, 1]$) maps every bounded subset $\mathbf{A_{30}} \subset \mathbf{A_3}$ into a set

$$\mathbf{A_{30}^*} := \mathbf{T_3}(\mathbf{A_{30}}) = \{\mathbf{T_3}[u, \mu] : u \in \mathbf{A_{30}}, \quad \mu \in [0, 1]\}$$

which is bounded in the space $C_S^{1+\beta}(\overline{D})$. From the suitable imbedding theorem the closure $\overline{\mathbf{A_{30}^*}}$ is a compact set in the space $C_S^{1+\alpha}(\overline{D})$ for $0 < \alpha < \beta < 1$. Hence, for any fixed μ, $\mathbf{T_3}[u, \mu]$ is a compact transformation.

v. From Theorem A.2 it follows that there exists a constant $M > 0$ such that every possible solution z of the equation $z = \mathbf{T_3}[z, \mu]$ ($z \in \mathbf{A_3}$, $\mu \in [0, 1]$), i.e., solution of the problem

$$\begin{cases} \mathcal{F}^i[z^i](t, x) = \mu\left(\mathbf{F}^i[z](t, x) - \mathcal{F}^i[\Phi^i](t, x)\right) + \mathcal{F}^i[\Phi^i](t, x), & i \in S, \quad \text{in } D, \\ z(t, x) = \phi(t, x) \quad \text{on } \Gamma, \end{cases}$$

for $z \in \mathbf{A_3}$, $\mu \in [0, 1]$, satisfies the estimate

$$\|z\|_{1+\alpha} \leq M.$$

vi. The equation $z = \mathbf{T_3}[z, 0]$ has the unique solution in the set $\mathbf{A_3}$ because by virtue of Theorems A.1 and A.2, the problem

$$\begin{cases} \mathcal{F}^i[z^i](t, x) = \mathcal{F}^i[\Phi^i](t, x), & i \in S, \quad \text{in } D, \\ z(t, x) = \phi(t, x) & \text{on } \Gamma, \end{cases}$$

has the unique solution in $\mathbf{A_3}$.

Finally, by (i)–(vi) and the Leray-Schauder theorem, in the set $\mathbf{A_3}$, there exists the solution z of the equation $z = \mathbf{T_3}[z, 1]$. Therefore, by the definition of the transformation $\mathbf{T_3}$, the function z is a solution of the considered problem

$$\begin{cases} \mathcal{F}^i[z^i](t, x) = \mathbf{F}^i[z](t, x), & i \in S, \quad \text{in } D, \\ z(t, x) = \phi(t, x) & \text{on } \Gamma, \end{cases}$$

in the space $C_S^{1+\beta}(\overline{D})$ and, moreover, $z \in C_S^{2+\alpha}(\overline{D})$ in the whole domain $\overline{D}$ in the space $C_S^{1+\beta}(\overline{D})$. This completes the proof. $\qquad\square$

CHAPTER 8

Stability of Solutions

8.1 Introduction

In this chapter, we will apply the monotone iterative method (more precisely: a certain variant of the Chaplygin method) to prove that problem (2.22) and (2.23) has at least one regular solution in a suitable class of functions and we will define a set in which we can expect a solution. We will also give some conditions which will guarantee the existence of a stable solution of problem (2.30)–(2.33) in $\overline{D}$.

We note that it is not possible to obtain all solutions by means of monotone iterative methods. For a large class of equations, there are solutions which cannot be included in the sector $\langle U_0, V_0 \rangle$ formed by the ordered pair of a lower solution U_0 and an upper solution V_0 of the problem (2.22) and (2.23).

Moreover, it is shown that by applying monotone iterative methods only stable solutions of the corresponding parabolic initial-boundary value problem can be obtained. Namely, we will prove that the limit of the solution of problem (2.30)–(2.33) as $t \to \infty$ is the solution of corresponding problem (2.22) and (2.23) obtained by the monotone iterative method.

This chapter focuses on infinite systems of equations.

8.2 Existence of Solutions for Infinite Systems

Let us consider the *Dirichlet problem* for an infinite system (2.22) with the boundary condition (2.23), i.e.,

$$\begin{cases} -\mathcal{A}^i[Z^i](x) = f^i(x, Z(x), Z) & \text{for } x \in G, \\ Z^i(x) = h(x) & \text{for } x \in \partial G \quad \text{and} \quad i \in S \end{cases}$$

and let U_0 and V_0 be an ordered pair of a lower and an upper solution of this problem in $\overline{G}$.

We assume that the functions

$$f^i : \overline{G} \times \mathcal{B}(S) \times L^p_S(G) \to \mathbb{R}, \quad (x, y, s) \mapsto f^i(x, y, s), \quad i \in S$$

Mathematical Neuroscience. http://dx.doi.org/10.1016/B978-0-12-411468-5.00008-9

satisfy the following assumptions in the set $\mathcal{K}_e$:

($\tilde{\mathbf{H}}_f$) $f^i(\cdot, y, s) \in C^{0+\alpha}(\overline{G})$ for $y \in \langle \underline{m}, \overline{M} \rangle$, $s \in \langle U_0, V_0 \rangle$, and $f^i(x, \cdot, \cdot)$ are continuous for $x \in \overline{G}$, $i \in S$.

($\tilde{\mathbf{W}}$) f^i, $i \in S$, satisfy the condition ($\mathbf{W}_+$) with respect to y and are increasing functions with respect to s (condition ($\mathbf{W}$)).

($\tilde{\mathbf{L}}$) f^i, $i \in S$, satisfy the Lipschitz condition with respect to y and s.

We assume that the operators $\mathcal{A}^i$, $i \in S$, are uniformly elliptic in $\overline{G}$, i.e., there exists a positive constant μ_0, such that

$$\sum_{j,k=1}^{m} a_{jk}^i(x)\xi_j\xi_k \geq \mu_0|\xi|^2 \quad \text{for all } \xi = (\xi_1, \ldots, \xi_m) \in \mathbb{R}^m \quad \text{and} \quad x \in \overline{G}, \quad i \in S,$$

where

($\tilde{\mathbf{H}}_a$) $a_{jk}^i \in C^{0+\alpha}(\overline{G})$, $a_{jk}^i = a_{kj}^i$ for $j, k = 1, \ldots, m$, $i \in S$. Moreover we assume that

($\tilde{\mathbf{H}}_h$) $h \in C_S^{2+\alpha}(\partial G)$.

Theorem 8.1 *Let assumptions* $\mathbf{A_0}$, ($\tilde{\mathbf{H}}_\mathbf{a}$), ($\tilde{\mathbf{H}}_\mathbf{f}$), ($\tilde{\mathbf{H}}_\mathbf{h}$) *hold, and conditions* ($\tilde{\mathbf{W}}$), ($\tilde{\mathbf{L}}$) *hold in the set* $\mathcal{K}_e$. *If the successive terms of the Chaplygin's sequences* $\{\mathring{U}_n\}$ *and* $\{\mathring{V}_n\}$, *where* $\mathring{U}_n = \{\mathring{U}_n^i\}_{i \in S}$, $\mathring{V}_n = \{\mathring{V}_n^i\}_{i \in S}$, $n = 1, 2, \ldots$, *that is the functions* $\mathring{U}_n^i$, $\mathring{V}_n^i$, $i \in S$, *for* $n = 1, 2, \ldots$ *are defined as regular solutions in* $\overline{G}$ *of the following infinite systems of equations*:

$$-(\mathcal{A}^i - \kappa\mathcal{I})[\mathring{U}_n^i](x) = f^i(x, \mathring{U}_{n-1}(x), \mathring{U}_{n-1}) + \kappa\mathring{U}_{n-1}^i(x), \tag{8.1}$$

$$-(\mathcal{A}^i - \kappa\mathcal{I})[\mathring{V}_n^i](x) = f^i(x, \mathring{V}_{n-1}(x), \mathring{V}_{n-1}) + \kappa\mathring{V}_{n-1}^i(x), \quad i \in S \tag{8.2}$$

for $n = 1, 2, \ldots$ *in G with boundary condition* (2.23) *and let an ordered pair of a lower solution* U_0 *and an upper solution* V_0 *of problem* (2.22) *and* (2.23) *in* $\overline{G}$ *(given by* $\mathbf{A_0}$*) be the pair of initial iterations in the iterative process, i.e.,* $\mathring{U}_0 = U_0$ *and* $\mathring{V}_0 = V_0$, *then*

i. $\{\mathring{U}_n\}$, $\{\mathring{V}_n\}$ *are well defined and* $\mathring{U}_n$, $\mathring{V}_n \in C_S^{2+\alpha}(\overline{G})$ *for* $n = 1, 2, \ldots$;

ii. *the inequalities*

$$U_0(x) \leq \mathring{U}_{n-1}(x) \leq \mathring{U}_n(x) \leq \mathring{V}_n(x) \leq \mathring{V}_{n-1}(x) \leq V_0(x), \quad n = 1, 2, \ldots \tag{8.3}$$

hold for $x \in \overline{G}$ *and the functions* $\mathring{U}_n$ *for* $n = 1, 2, \ldots$, *are lower solutions of problem* (2.22) *and* (2.23) *in* $\overline{G}$, *and analogously,* $\mathring{V}_n$ *for* $n = 1, 2, \ldots$, *are upper solutions of this problem*;

iii. *the functions* $\underline{Z} = \{\underline{Z}^i\}_{i \in S}$ *and* $\overline{Z} = \{\overline{Z}^i\}_{i \in S}$, *where*

$$\underline{Z}^i(x) := \lim_{n \to \infty} \mathring{U}_n^i(x), \quad \overline{Z}^i(x) := \lim_{n \to \infty} \mathring{V}_n^i(x), \quad i \in S, \quad \text{for } x \in \overline{G} \qquad (8.4)$$

are the minimal and maximal regular solutions of problem (2.22) *and* (2.23) *in* $\overline{G}$,

$$U_0(x) \le \underline{Z}(x) \le \overline{Z}(x) \le V_0(x) \quad \text{for } x \in \overline{G} \qquad (8.5)$$

and $\underline{Z}, \overline{Z} \in C_S^{2+\alpha}(\overline{G})$.

Remark 8.1 We remark that problem (8.1), (2.23) and (8.2), (2.23) is a system of separate problems, each with only one equation. Therefore, its solution is a collection of separate solutions of these problems.

8.3 Stability of Solutions of Infinite Systems

Let us consider the initial-boundary value problem

$$\begin{cases} \mathcal{D}_t z^i(t, x) - \mathcal{A}^i[z^i](t, x) - f^i(x, z(t, x), z(t, \cdot)) = 0 & \text{for } (t, x) \in D, \\ z^i(0, x) = \phi_0^i(x) & \text{for } x \in G, \\ z^i(t, x) = 0 & \text{for } (t, x) \in \sigma \text{ and } i \in S \end{cases} \qquad (8.6)$$

with the compatibility condition $\phi_0(x) = 0$ for $x \in \partial G$.

We have the following Theorem:

Theorem 8.2 *Let assumptions* $\mathbf{A_0}$, $(\tilde{\mathbf{H}}_a)$, $(\tilde{\mathbf{H}}_\phi)$, $(\tilde{\mathbf{H}}_f)$ *hold, and conditions* $(\tilde{\mathbf{W}})$, $(\tilde{\mathbf{L}})$ *hold in the set* $\mathcal{K}$. *Then problem* (8.6) *has the unique global regular solution* $z = z(t, x)$ *within the sector* $\langle u_0, v_0 \rangle$, *and* $z \in C_S^{2+\alpha}(\overline{D})$.

The proof is simple and follows from the same arguments as that for the Theorem 2.1 so we may omit it here.

In addition to this problem we consider the corresponding problem

$$\begin{cases} -\mathcal{A}^i[Z^i](x) - f^i(x, Z(x), Z(\cdot)) = 0 & \text{for } x \in G, \\ Z^i(x) = 0 & \text{for } x \in \partial G \text{ and } i \in S. \end{cases} \qquad (8.7)$$

Theorem 8.3 *Let assumptions* $(\mathbf{H}_a)$, $(\mathbf{H}_f)$ *hold, and conditions* $(\tilde{\mathbf{W}})$, $(\tilde{\mathbf{L}})$ *hold in the set* $\mathcal{K}_e$. *Let* $V_0 = V_0(x)$ *be an upper solution of problem* (8.7) *in* $\overline{G}$, $V_0 \in C_S^{2+\alpha}(\overline{G})$ *and let* $\tilde{v} = \tilde{v}(t, x)$ *be a solution of the initial-boundary value problem when* $\phi_0(x) = V_0(x)$ *for* $x \in \overline{G}$, *i.e.*,

$$\begin{cases} \mathcal{D}_t \tilde{v}^i(t, x) - \mathcal{A}^i[\tilde{v}^i](t, x) - f^i(x, \tilde{v}(t, x), \tilde{v}(t, \cdot)) = 0, & i \in S, \quad \text{for } (t, x) \in D, \\ \tilde{v}^i(0, x) = V_0^i(x) & \text{for } x \in G, \\ \tilde{v}^i(t, x) = 0 & \text{for } (t, x) \in \sigma \end{cases} \qquad (8.8)$$

with the compatibility condition $V_0(x) = 0$ for $x \in \partial G$. Then

$$\frac{\partial \tilde{v}}{\partial t} \le 0 \quad \text{and} \quad \tilde{v}(t, x) \le \tilde{v}(0, x) = V_0(x) \quad \text{in } \overline{D}.$$

Proof. Under the above assumptions, it follows that problem (8.8) has the unique regular solution $\tilde{v} = \tilde{v}(t, x)$ in $\overline{D}$.

On the other hand, V_0 is an upper solution for problem (8.7). Since $\mathcal{D}_t V_0^i(x) = 0$, V_0 is a solution of the problem

$$\begin{cases} \mathcal{D}_t V_0^i(x) - \mathcal{A}^i[V_0^i](x) - f(x, V_0(x), V_0(\cdot)) = 0 & \text{for } (t, x) \in D, \\ V_0^i(x) = V_0^i(x) & \text{for } x \in G, \\ V_0^i(x) \ge 0 & \text{for } (t, x) \in \sigma \text{ and } i \in S. \end{cases} \tag{8.9}$$

Applying Szarski's theorem on weak differential inequalities (Theorem 3.5) to systems (8.8) and (8.9) we get

$$\tilde{v}(t, x) \le V_0(x) \quad \text{in } \overline{D}.$$

Now we consider the function

$$\tilde{v}_\kappa(t, x) := \tilde{v}(t + \kappa, x) \quad \text{for } \kappa > 0.$$

This function satisfies the following problem:

$$\begin{cases} \mathcal{D}_t \tilde{v}_\kappa^i(t, x) - \mathcal{A}^i[\tilde{v}_\kappa^i](t, x) - f^i(x, \tilde{v}_\kappa(t, x), \tilde{v}_\kappa(t, \cdot)) = 0 & \text{for } (t, x) \in D, \\ \tilde{v}_\kappa^i(0, x) = \tilde{v}^i(0 + \kappa, x) = \tilde{v}^i(\kappa, x) \le V_0^i(x) & \text{for } x \in G, \\ \tilde{v}_\kappa^i(t, x) = 0 & \text{for } (t, x) \in \sigma \text{ and } i \in S. \end{cases} \tag{8.10}$$

Applying again Theorem 3.5 to systems (8.8) and (8.10) we get

$$\tilde{v}_\kappa(t, x) \le \tilde{v}(t, x) \quad \text{in } \overline{D}.$$

The function $\tilde{v}(t, x)$ is nonincreasing with respect to t. Indeed, let $0 \le t_1 < t_2$ and $\kappa = t_2 - t_1$. Then

$$\tilde{v}(t_1, x) \ge \tilde{v}_\kappa(t_1, x) = \tilde{v}(t_1 + \kappa, x) = \tilde{v}(t_2, x).$$

This completes the proof. □

We can prove the analogous theorem for a lower solution $U_0 = U_0(x)$ of the problem (8.8).

Theorem 8.4 *Let assumptions $\mathbf{A_0}$, $(\mathbf{H_a})$, $(\tilde{\mathbf{H}}_\mathbf{f})$, $(\tilde{\mathbf{H}}_\phi)$ hold, and conditions $(\tilde{\mathbf{W}})$, $(\tilde{\mathbf{L}})$ hold in the set $\mathcal{K}$. If $u = u(t, x)$ is a regular uniformly bounded solution of the initial-boundary value problem*

$$\begin{cases} \mathcal{D}_t u^i(t, x) - \mathcal{A}^i[u^i](t, x) - f^i(x, u(t, x), u(t, \cdot)) = 0 & \text{for } (t, x) \in D, \\ u^i(0, x) = \phi_0^i(x) & \text{for } x \in G, \\ u^i(t, x) = 0 & \text{for } t > 0, \quad x \in \partial G \text{ and } i \in S, \end{cases} \tag{8.11}$$

when $\phi_0 \in \langle U_0, V_0 \rangle$ *and* $\phi_0(x) = 0$ *for* $x \in \partial G$, *and* $\lim_{t \to \infty} u(t, x) = \hat{U}(x)$ *exists, then the function* $\hat{U} = \hat{U}(x)$ *is a regular solution of the corresponding boundary value problem*

$$\begin{cases} -\mathcal{A}^i[\hat{U}^i](x) - f^i(x, \hat{U}(x), \hat{U}(\cdot)) = 0 & \text{for } x \in G, \\ \hat{U}^i(x) = 0 & \text{for } x \in \partial G \ \text{and} \ i \in S. \end{cases} \tag{8.12}$$

Theorem 8.5 *Let assumptions of Theorem 8.4 hold. Let* V_0 *be an upper solution and* $\overline{Z}$ *be the maximal regular solution of problem (8.7) in* $\overline{G}$ *defined by (8.4). If* $u = u(t, x)$ *is a solution of problem (8.6) in* $\overline{D}$ *with initial condition* φ_0 *such that*

$$\overline{Z}(x) \preceq \varphi_0(x) \preceq V_0(x) \quad \text{in } \overline{G},$$

then $\lim_{t \to \infty} u(t, x) = \overline{Z}(x)$, *so* $\overline{Z}$ *is asymptotically stable from above.*

Let U_0 *be a lower solution and* $\underline{Z}$ *be the minimal regular solution of problem (8.7) in* $\overline{G}$ *defined by (8.4). If* $u = u(t, x)$ *is a solution of problem (8.6) in* $\overline{D}$ *with initial condition* φ_0 *such that*

$$U_0(x) \leq \varphi_0(x) \leq \underline{Z}(x) \quad \text{in } \overline{G},$$

then $\lim_{t \to \infty} u(t, x) = \underline{Z}(x)$, *so* $\underline{Z}$ *is asymptotically stable from below.*

If $\overline{Z} = \underline{Z} := Z$, *i.e., regular problem (8.7) has the unique solution* Z *and*

$$U_0(x) \leq \varphi_0(x) \leq V_0(x) \quad \text{in } \overline{G},$$

then $\lim_{t \to \infty} u(t, x) = Z(x)$, *so* Z *is asymptotically stable, i.e., both from above and from below.*

Proof. From the Theorems 4.1 and 8.1, each solution $u = u(t, x)$ of problem (8.6) with the initial inequalities

$$U_0 \leq \varphi_0(x) \leq V_0(x) \text{ in } \overline{G}$$

satisfies

$$U_0(x) \leq \tilde{u}(t, x) \leq u(t, x) \leq \tilde{v}(t, x) \leq V_0(x) \text{ in } \overline{D}$$

By assumption, the solution $\tilde{v}$ satisfies

$$\overline{Z}(x) \leq \tilde{v}(t, x) \leq V_0(x) \text{ in } \overline{D}$$

Hence, $\tilde{v}$ is bounded from below and by virtue of Theorem 8.3, the function $\tilde{v}$ is non-decreasing with respect to t. Therefore, $\lim_{t \to \infty} \tilde{v}(t, x)$ exists. By Theorem 8.4, this limit is a solution of problem (8.7) and $\overline{Z} \leq \lim_{t \to \infty} \tilde{v}(t, x)$.

On the other hand, Z is the maximal solution of (8.7), so we get

$$\lim_{t \to \infty} \tilde{v}(t, x) = \overline{Z}(x). \tag{8.13}$$

Any solution $u = u(t, x)$ of problem (8.6) in $\overline{D}$ with an initial condition φ_0 satisfying the inequalities

$$\overline{Z}(x) \leq \varphi_0(x) \leq V_0(x) \text{ in } \overline{G}$$

satisfies

$$\overline{Z}(x) \leq u(t, x) \leq \tilde{v}(t, x) \text{ in } \overline{D}$$

Moreover, (8.13) holds and hence $\lim_{t \to \infty} u(t, x) = \overline{Z}(x)$, so $\overline{Z}$ is an asymptotically stable solution from above of problem (8.7). The rest of the proof runs analogously. $\quad\square$

Part II

Application of Nonlinear Analysis

CHAPTER 9

Introduction to Part II

In computational neuroscience there is relatively little quality work done by mathematical neuroscientists. At present, computational neuroscience researchers assume a direct causal relationship between a mechanism and a phenomenon, ignoring the constraints that higher-level properties exert on the possible functions of that mechanism. One of these constraints is dynamic continuity which is intrinsically difficult to harness approximately since it is subject to dynamical misalignment, producing a false sense of reality. Under these circumstances, methods of approximation that are often used can delude dynamic continuity inherent in the brain's integrative operation.

It is clear that techniques from nonlinear dynamical systems theory and mathematical physics have proven useful to date with sophisticated tools of modern applied mathematics, including nonlinear waves and bifurcation analysis, the use of geometric singular perturbation theory, and numerical methods. However, such methods fall short of unifying the divisions in the brain's integrative operation seamlessly through dynamic continuity. Harnessing continuity in the dynamics of the brain's integrative operation at different levels of neuronal organization requires the use of modern mathematics as a means of representing infinite systems of equations in a vector space of continuous functions. Continuous functions represented by these infinite systems of equations of reaction-diffusion type in partially ordered vector spaces represent hierarchical levels in a way that will not delude the continuous dynamics of the neuronal systems across spatiotemporal scales through approximations. A vector space can be thought of as a nonempty set from which has been swept away all structure irrelevant to the continuity of functions defined on it. Continuous functions are the chief objects of interest and vector spaces are regarded as carriers of such functions and as domains over which they can be integrated. These ideas lead naturally into the theory of Banach space, which is an infinite-dimensional vector space that is connected. Such connected spaces provide the ideal context for dynamic continuity which is useful in applications to neuroscience.

This book highlights attempts to include a mathematical description of dynamic continuity across scale from the cellular to the systems level. The epitomization of the process starts not from the mathematical description of neurobiological processes but from the mathematical description of a more general class of dynamical systems for which models of neural structures are a special case. It is therefore unique since Part I presents a mathematical theory of infinite systems of partial differential equations based on methods of nonlinear functional analysis and Part II applies such infinite systems in neuroscience. Why are infinite systems so important? The brain is considered dynamically an open system, where information is unlimited and there are an almost infinite number of unique possibilities for encoding of this information through physical interactions. This requires an infinite system of equations to represent the repertoire of possibilities for encoding of information. Moreover, the dynamic complexity of the brain at various levels of investigation as depicted in Figure 9.1 is a continuum. The dynamics in order to unify and explain the available behavioral and physiological data across each hierarchical level of functional organization should be continuous. Infinite systems of partial differential equations are continuous dynamical representations. Infinite systems allow for integration to take place continuously due to the fact that the number of both equations and variables may be infinite, so that processes of a particular neural system at each hierarchical level can be represented as part of a continuum.

Integration in neuroscience requires synthesis of dynamic continuity, so almost all phenomena modeled must be done with continuous functions defined on infinite-dimensional vector spaces that consist of partial differential equations of the reaction-diffusion type. The non space-clamped Hodgkin-Huxley equations are an infinite-dimensional system of nonlinear reaction-diffusion equations:

$$C_m \frac{\partial V}{\partial t} = \frac{a}{2\rho} \frac{\partial^2 V}{\partial x^2} - \sum g_i(x, t; V)(V - V_i),$$

where $V(x, t)$ is the membrane potential (mV) and $g_i(x, t; V)$ is the conductance change and is dependent on the membrane potential of the neuron. The Hodgkin-Huxley equations for the

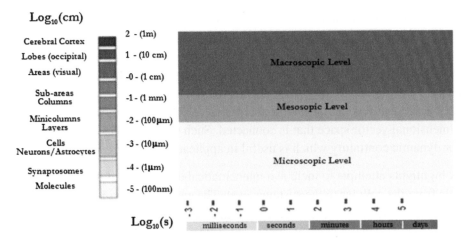

Figure 9.1 Illustrating the hierarchy of spatiotemporal evolution. Please see color plate at the back of the book.

excitable membrane of the squid giant axon are:

$$\sum g_i(x, t; V)(V - V_i) = m^3 g_{Na} h(V - V_{Na}) + n^4 g_K(V - V_K) + g_L(V - V_L),$$

where gating variables m, n, and h obey first-order kinetics with voltage-dependent rate coefficients α, β. The essential part of the Hodgkin-Huxley model is its nonlinearity. In essence, once the voltage changes, the ionic conductances change to induce even more changes in the voltage. Threshold properties ensure that the behavior of the solution $V(x, t) \to 0$ as $t \to \infty$ if $V(x, 0)$ is sufficiently small (subthreshold) and $V(x, t) \to \infty$ as $t \to \infty$ if $V(x, 0)$ is sufficiently large (suprathreshold). Unfortunately, in mathematical neuroscience the complete Hodgkin-Huxley system is too complex for analysis and most often caricatures are taken to reproduce qualitatively the behavior of the solutions for the propagation of a nerve impulse (or spike) down a cable structure. Typical examples are the FitzHugh-Nagumo equations:

$$C_m \frac{\partial V}{\partial t} + \omega = \frac{a}{2\rho} \frac{\partial^2 V}{\partial x^2} + f(V),$$

$$\frac{\partial \omega}{\partial t} = \sigma V - \gamma \omega,$$

where ω is a recovery variable that ensures a pulse solution is reproduced and $f(V)$ is a nonlinear functional describing the internal diffusion of membrane potential in a cable whose membrane current is given by a cubic equation. The use of the cubic polynomials to represent the membrane current-voltage relation was attributed to Andrew Huxley and used by Richard FitzHugh. The N-shaped dendrite used in nonlinear cable theory developed by Aron Gutman is often associated with the use of a piecewise linear cubic equation. The FitzHugh-Nagumo model has been used in mathematical neuroscience to explore threshold behavior, action potentials (traveling wave solutions), and repetitive activity (time-periodic solutions). When the recovery variable is held fixed at its steady-state, the Hodgkin-Huxley model is bistable and when $\omega = 0$ the FitzHugh-Nagumo system reduces to the so-called Nagumo's equation:

$$C_m \frac{\partial V}{\partial t} = \frac{a}{2\rho} \frac{\partial^2 V}{\partial x^2} + f(V).$$

Specifically, this equation is a nonlinear diffusion equation whose behavior of solution expresses leading edge fronts and not pulse solutions.

Computational neuroscience is founded on the assumption that decomposition of neuronal systems does not change the behavior of solutions. This has been referred to as the "lumped parameter assumption." Consequently this assumption emits a way of decomposing excitable systems. An excitable medium is considered to be an infinite-dimensional system of partial differential equations. Although the excitation equations that represent neuronal models are infinite-dimensional and nonlinear, can their solutions equally well be reproduced by finite-dimensional systems? This is the motivating question that we have answered using modern functional analysis methods. The answer is that the solutions are not usually equally well reproduced.

In Part II we examine the validity of the "lumped parameter assumption" by focusing on the structure of neuronal models when a continuously distributed neuron is divided into sufficiently small segments (or compartments). If the assumption is true then one makes a negligible small error by assuming that each compartment is isopotential and uniform in its properties. From this perspective spanned a plethora of so-called spiking neuron models consisting of single compartments for each neuron with a number of ionic channels modeled using the spaced-clamped Hodgkin and Huxley formalism. In consequence of this assumption, the behavior of solution of the spaced-clamped version of the Hodgkin and Huxley equations is assumed to be representative of the behavior of solutions expected of the entire neuron. Thus the state-space of the Hodgkin-Huxley system is four-dimensional (V, m, n, h):

$$C_m \frac{dV}{dt} = \sum g_i(x, t; V)(V - V_i),$$

$$\frac{dm}{dt} = \alpha_m(1 - m) - \beta_m m,$$

$$\frac{dn}{dt} = \alpha_n(1 - n) - \beta_n n,$$

$$\frac{dh}{dt} = \alpha_h(1 - h) - \beta_h h.$$

At this point, the propagating spike corresponds to a traveling wave solution at a constant speed (v). The analysis suggests that the similarity transformation $\xi = x + vt$ reduces the behavior of the solutions of the propagating spike exhibited by an infinite system to that of a differential system dependent only on the new traveling wave variable (ξ). Back in the 1970s, there was significant progress in using comparison and maximum principles to rigorously analyze the existence, uniqueness, and stability properties of traveling wave solutions to reaction-diffusion equations. However, the analysis did show that the "lumped parameter assumption" applied in relatively few cases and in limited conditions. For instance, James Keener had shown that the spatially discrete analog to the reduced FitzHugh-Nagumo equation or Nagumo's equation:

$$C_m \frac{dV_n}{dt} = \frac{a}{2\rho h^2}(V_{n+1} - 2V_n + V_{n-1}) + f(v_n)$$

has traveling wave solutions (i.e., wave fronts) and the speed of propagation approaches the speed of continuous model when $h \ll 1$ otherwise propagation fails.

The conceptual framework of Part II is infinite-dimensional dynamical systems. A dynamical system is any system that is realized by an input-output function. By a realization of an input-output function we mean any dynamical system when subjected to the same input will produce the same output as the experimental data. In this sense, a dynamical system goes beyond explaining experimental data in dynamical terms, but also surveys all possible experimental predictions. That is why mathematical neuroscience is essential and the essence stems from the fact that models arising in the neurosciences require a conceptual framework beyond computational solutions of the highly nonlinear equations due to the limited validity of the "lumped parameter assumption" used in computational neuroscience. Bifurcation theory and phase-plane methods are important so-called geometrical methods used in the study of dynamical systems represented by ordinary differential equations, yet are limited because of the inherent acceptance of the "lumped parameter assumption." Moreover, for sufficiently localized initial conditions, if the asymptotic solutions determined by linearizing about the equilibrium point are sensitive to perturbations then this means that standard perturbation methods break down. Neither approximate nor computational solutions can be used and the new analytical approaches are needed. Nonlinear dynamics attained using geometrical and computational methods for ordinary differential equations often exhibit limited behaviors of the solutions than those expressed by infinite-dimensional dynamical systems represented by partial differential equations.

In Chapter 10 we have gone beyond existing literature by including a much larger proportion of theory and methods of nonlinear analysis with applications to neuronal models in infinite-dimensional vector spaces. Methods of functional analysis for the nonlinear analysis of infinite systems have important applications in neuroscience and must not be limited to just bifurcation analysis, simply because the latter is cumbersome and not well developed for infinite-dimensional vector spaces.

In Chapter 11 cable equations are analyzed to describe the invalidity of the "lumped parameter assumption" particularly used in compartmental models of neurons. Nonlinear analysis using comparison theorems and maximum principles, and the Gronwall-Bellman lemma, compare the validity of the "lumped parameter assumption" in describing the behavior of solutions of cable equation. Thus by using examples from equations exhibited by neuronal cable theory, the nonlinear analysis confirms that compartmental models, including various spiking neuron models, are poor quantifiers of the dynamics exhibited by neurons. In particular, the cable model represented by a nonlinear cable equation has different solutions from those of the compartmental model. Specifically, we show that these solutions converge in the limit as $t \to \infty$; in other words, the compartment model is only a valid approximation to the cable equation in the steady-state. It is shown that the steady-state solution is stable and this stability applies to the transient problem. That is to explain that the solutions are different while their stability is inferred from the methods implied from the steady-state solution. Furthermore, unlike axonal spikes, dendritic spikes attenuate and propagate with a nonconstant speed. Thus, the mathematical analysis of the nonlinear cable equation for active dendrites whose behavior of solutions may differ from those derived from the Hodgkin-Huxley equations is a testament to the confusion that solutions of the Hodgkin-Huxley equations underlie our understanding of how signals are propagated in neurons.

Chapter 12 explores comparison theorems for finite and infinite reaction-diffusion systems; asymptotic behavior of the solutions; theorems on monotonous convergence of a solution of a transient problem with $t \to \infty$; provided that the limit is a solution to the problem; stability analysis for finite systems; proving that a solution to a stationary problem is a stable solution to a transient problem; determining invariant sets.

Part II attempts at re-shaping mathematical neuroscience to enable mathematicians to follow the path in developing a well-grounded theoretical neuroscience.

CHAPTER 10

Continuous and Discrete Models of Neural Systems

10.1 Introduction

The application of systems of nonlinear partial differential equations to describe the dynamics of neural systems assumes that the number of variables involved in the modeling processes is unbounded. We are adopting an important assumption that the numbers of both equations and variables may be infinite. This assumption, in turn, leads to models involving infinite systems of equations. While constructing models, there appear the following two main descriptions of the processes considered: a discrete model and a continuous model. If a variable taking a countable infinite number of values is used to describe the process then a discrete model of this process is obtained. Discrete models are expressed in terms of infinite countable systems of equations. On the other hand, if continuous space and time variables are used then a continuous model is obtained which is expressed in terms of infinite uncountable systems of equations.

An elementary explanation may be given as to why two different (discrete and continuous) models emerge in the description of processes which arise in neuroscience. If a variable $z^i = z^i(t, x)$ is a discrete one, taking nonnegative integer values, the model thus obtained is a discrete model built of infinite countable systems of equations. On the other hand, when a continuous variable taking nonnegative real values is used in the description, we arrive at a continuous model built of infinite uncountable systems of equations. This means that the study of continuous models of neural systems reduces to the study of infinite uncountable systems of equations describing the model considered.

10.2 Mathematical Motivations

Now we are going to present two examples of FitzHugh-Nagumo partial differential equations which lead to infinite uncountable systems of equations. In other words, we are going to give motivation for and the source of the study of infinite uncountable systems of equations.

Mathematical Neuroscience. http://dx.doi.org/10.1016/B978-0-12-411468-5.00010-7

Example Consider the Cauchy problem

$$\begin{cases} \dfrac{\partial u(t,x)}{\partial t} = d\dfrac{\partial^2 u(t,x)}{\partial x^2} + f(t,x,u(t,x)) & \text{for } (t,x) \in (0,T] \times \mathbb{R}^m := D, \\ u(0,x) = \varphi(x) & \text{for } x \in \mathbb{R}^m, \end{cases} \tag{10.1}$$

where the diffusion coefficient $d \geq 0$ and f is continuous.

If the diffusion coefficient d is equal to zero, then the corresponding equation is an ordinary differential equation with x as a parameter or infinite uncountable system of equations

$$\begin{cases} \dfrac{du(t,x)}{dt} = f(t,x,u(t,x)) & \text{for } t \in (0,T], \\ u(0,x) = \varphi(x) & \text{for } x \in \mathbb{R}^m. \end{cases} \tag{10.2}$$

There is an important example with zero diffusion such as the (parabolic) FitzHugh-Nagumo equation for the conduction of nerve impulses.

Example Consider the FitzHugh-Nagumo system of equations of two dependent variables

$$\begin{cases} \dfrac{\partial u(t,x)}{\partial t} = d_1 \dfrac{\partial^2 u(t,x)}{\partial x^2} + \alpha + u(t,x) + v(t,x) - \dfrac{1}{3}u^3(t,x), \\ \dfrac{\partial v(t,x)}{\partial t} = d_2 \dfrac{\partial^2 v(t,x)}{\partial x^2} - \dfrac{1}{2}u(t,x) - \dfrac{1}{4}v(t,x) & \text{for } (t,x) \in (0,T] \times \langle 0,1 \rangle, \end{cases} \tag{10.3}$$

where diffusion coefficients d_1, d_2 are nonnegative $d_1 \geq 0, d_2 \geq 0$ with homogeneous Neumann boundary conditions,

$$\frac{\partial u(t,0)}{\partial x} = \frac{\partial v(t,0)}{\partial x} = 0,$$

$$\frac{\partial u(t,1)}{\partial x} = \frac{\partial v(t,1)}{\partial x} = 0 \quad \text{for } t \geq 0,$$

and initial conditions

$$u(0,x) = u_0(x),$$
$$v(0,x) = v_0(x) \quad \text{for } x \in \langle 0,1 \rangle.$$

If the diffusion coefficients are equal to 0, then the corresponding system of ordinary differential equations is of the form

$$\begin{cases} \dfrac{du(t,x)}{dt} = \alpha + u(t,x) + v(t,x) - \dfrac{1}{3}u^3(t,x), \\ \dfrac{dv(t,x)}{dt} = -\dfrac{1}{2}u(t,x)) - \dfrac{1}{4}v(t,x) & \text{for } t \geq 0, \end{cases} \tag{10.4}$$

where $x \in \langle 0,1 \rangle$ is a parameter. This is an infinite uncountable system of equations.

10.3 The Formulation of Problems

We consider the infinite system of equations with functionals of the form:

$$\mathcal{F}_c^i[z^i](t, x) := \frac{\partial z^i(t, x)}{\partial t} - \mathcal{L}_c^i[z^i](t, x) = f^i(t, x, z(t, x), z) \tag{10.5}$$

for $(t, x) \in D$, and $i \in S$, where $x = (x_1, x_2, \ldots, x_m)$, $(t, x) \in (0, T] \times G := D$, $0 < T < \infty$, $G \subset \mathbb{R}^m$, G is an open and bounded domain whose boundary ∂G is an $(m-1)$-dimensional surface of a class $C^{2+\alpha}$ for some α, $0 < \alpha \leq 1$, S is an arbitrary infinite (countable or uncountable) set of indices and differential operators

$$\mathcal{F}_c^i := \frac{\partial}{\partial t} - \mathcal{L}_c^i, \quad \mathcal{L}_c^i := \sum_{k,l=1}^{m} a_{kl}^i(t, x)\frac{\partial^2}{\partial x_k \partial x_l} - \sum_{k=1}^{m} b_k^i(t, x)\frac{\partial}{\partial x_k} - c^i(t, x)\mathcal{I}, \quad i \in S \tag{10.6}$$

are diagonal operators,[1] with the homogeneous initial and boundary conditions

$$\begin{aligned} z^i(0, x) &= 0 \quad \text{for } x \in G, \\ z^i(t, x) &= 0 \quad \text{for } (t, x) \in \sigma := (0, T) \times \partial G \quad \text{and} \quad i \in S \end{aligned} \tag{10.7}$$

where $\mathcal{I}$ is the identity operator, $z = z(\cdot, \cdot)$ and $z = \{z^i\}_{i \in S}$, $z^i = z^i(t, x)$ stands for the mapping

$$z : \overline{D} \times S \mapsto \mathbb{R}, \quad (t, x, i) \to z(t, x, i) := z^i(t, x)$$

and f^i, $i \in S$ are given functions

$$f^i : \overline{D} \times \mathcal{B}(S) \times C_S(\overline{D}) \to \mathbb{R}, \quad (t, x, y, s) \mapsto f^i(t, x, y, s), \quad i \in S.$$

We assume that system (10.5) is uniformly parabolic with respect to $(t, x, i) \in \overline{D} \times S$.

10.4 Observations

Observation 10.1 To prove the existence (and uniqueness) of solution of problem (10.5) and (10.7), we shall apply a fixed point theorem for contraction mappings (or compact mappings) in suitably selected Banach space $\mathcal{X}$ (e.g., Hölder space $C_S^{1+\alpha}(\overline{D})$ or $C_S^{2+\alpha}(\overline{D})$).

Finding a solution to problem (10.5), (10.7) reduces to the identification of a fixed point of the appropriate transformation T in a certain subset of the space $\mathcal{X}$.

---

[1] The operator $\mathcal{L}_c := \{\mathcal{L}_c^i\}_{i \in S}$ is called diagonal if $\mathcal{L}_c^i$ depends on z^i only for all $i \in S$.

Therefore, we consider the closed subset **A** of the space $\mathcal{X}$ defined as follows:

$$\mathbf{A} := \{w \colon w \in \chi, w(t, x) = 0 \quad \text{on } \Gamma, \ (t, x) \in D\}, \tag{10.8}$$

and for $u \in \mathbf{A}$ we define the transformation **T** setting

$$\mathbf{T} \colon \mathbf{A} \to \mathbf{A}, \quad u \mapsto \mathbf{T}[u] := z, \tag{10.9}$$

where z is the (supposedly unique) solution of the following infinite system of equations:

$$\mathcal{F}_c^i[z^i](t, x) = f^i(t, x, u(t, x), u) := \mathbf{F}^i[u](t, x) \quad \text{for } i \in S \text{ in } D \tag{10.10}$$

with the homogeneous initial and boundary conditions

$$\begin{aligned} z^i(0, x) &= 0 \quad \text{for } x \in G, \\ z^i(t, x) &= 0 \quad \text{for } (t, x) \in \sigma \quad \text{and} \quad i \in S, \end{aligned} \tag{10.11}$$

where S is an arbitrary infinite countable or infinite uncountable set of indices, $f^i = f^i(t, x, y, s)$, $i \in S$, are given functions and $\mathbf{F} = \{\mathbf{F}^i\}_{i \in S}$ is the nonlinear operator generated by the functions $f^i, i \in S$.

The nonlinear operator $\mathbf{F} = \{\mathbf{F}^i\}_{i \in S}$ is generated by the functions $f^i(t, x, y, s)$, $i \in S$, and defined for sufficiently regular functions $\beta = \beta(t, x)$ as follows:

$$\mathbf{F} \colon \beta \mapsto \mathbf{F}[\beta],$$

where

$$\mathbf{F}^i[\beta^i](t, x) := f^i(t, x, \beta(t, x), \beta), \quad i \in S. \tag{10.12}$$

The nonlinear operator is sometimes also called the superposition operator, composition operator, or substitution operator. This type of operator plays an important role in the theory of nonlinear equations. The introduction of the nonlinear operator enables proofs of certain theorems to be simplified significantly and renders the proofs elegant.

Observe that system (10.10) has the following crucial property: for an arbitrary $i \in S$ the ith equation of this infinite system of equations depends on the ith unknown function $z^i = z^i(t, x)$ only. Therefore, this system is a collection of several equations each of which is an equation in one unknown function only.

Two situations are possible:

1. If system (10.10) is infinite countable, then $\overline{\overline{S}} = \overline{\overline{\mathbb{N}}} = \aleph_0.$[2] In this system for an arbitrary fixed index $j \in \mathbb{N}$, the jth equation depends on the one unknown function z^j where

---

[2] We use $\overline{\overline{S}}$ to denote the cardinality of the set S. The symbol $\aleph_0$ denotes the cardinality of the set of nonnegative integers; that is $\overline{\overline{\mathbb{N}}} = \aleph_0$. The symbol $\aleph$ (aleph) is the first letter of the Hebrew alphabet. The symbol $\aleph_0$ is

$z = \{z^j\}_{j \in \mathbb{N}}$ and $z^j = z^j(t, x) := z(t, x, j)$, only. In this case we consider the infinite countable system of equations of the form

$$\mathcal{F}_c^j[z^j](t, x) = f^j(t, x, u(t, x), u(\cdot, \cdot)) := \mathbf{F}^j[u](t, x) \quad \text{for } j \in \mathbb{N} \text{ in } D, \qquad (10.13)$$

which is a collection of the infinite countable number of separate scalar problems. These infinite countable systems appear as discrete models of processes considered.

2. Analogously, if system (10.10) is infinite uncountable, then $\overline{\overline{S}} = \overline{\overline{\mathbb{R}}} = \mathfrak{c}$. Let $\Lambda \subset \mathbb{R}^+$ be the compact set of indices and $\overline{\overline{\Lambda}} = \mathfrak{c}$, e.g., $\Lambda := [\lambda_0, \lambda_1] \subset \mathbb{R}^+$. In this system for an arbitrary fixed parameter $\lambda \in \Lambda$, the λth equation depends on the one unknown function z^λ where $z = \{z^\lambda\}_{\lambda \in \Lambda}$ and $z^\lambda = z^\lambda(t, x) := z(t, x, \lambda)$, only.

In this case we consider the infinite uncountable system of equations of the form

$$\mathcal{F}_c^\lambda[z^\lambda](t, x) = f(t, x, u(t, x), u(\cdot, \cdot)) := \mathbf{F}[u](t, x) \quad \text{for } \lambda \in \Lambda \text{ in } D, \qquad (10.14)$$

which is a collection of separate equations of the infinite uncountable number of scalar problems.

These infinite uncountable systems are continuous models of processes described. Therefore, the solutions of problem (10.13) and (10.11) and the solutions of problem (10.14) and (10.11) are collections of separate solutions of these infinite countable quantity scalar problems (10.13) and (10.11) or an infinite uncountable number of scalar problems (10.14) and (10.11). This means that in the case of infinite systems of equations (both infinite countable and infinite uncountable systems) can be examined with use of the topological fixed point method. Moreover, owing to such an approach, the countable and uncountable cases may be considered analogously and simultaneously.

Observation 10.2 We encounter a similar situation, while—using an approximation method—we are looking for an exact solution to problem (10.5) and (10.7) and to this end we construct sequences of consecutive approximation.

We are now going to consider the method of lower and upper solutions and to present the construction of sequences of consecutive approximations: a lower solution sequence $\{u_n^i(t, x)\}_{i \in S}$ and an upper solution sequence $\{v_n^i(t, x)\}_{i \in S}$. We start the iterative process from the zero iterations, that is functions u_0 and v_0, which are a lower and upper solution, respectively, for the problem considered (these functions have to be predefined).

---

pronounced "aleph zero." An infinite set of cardinality $\aleph_0$ is called infinite countable. The cardinality of the set of reals is denoted by $\mathfrak{c}$; that is $\overline{\overline{\mathbb{R}}} = \mathfrak{c}$. A set of cardinality $\mathfrak{c}$ is called a set of continuum cardinality. It is in particular an infinite uncountable set.

Relevant terms of the approximation sequences are defined as solutions to homogeneous initial-boundary value problems for infinite systems of equations of the form:

$$\begin{cases} \mathcal{F}_c^i[u_n^i](t, x) = f^i(t, x, u_{n-1}(t, x), u_{n-1}) := \mathbf{F}^i[u_{n-1}](t, x) & \text{for } (t, x) \in D, \\ u_n^i(0, x) = 0 & \text{for } x \in G, \\ u_n^i(t, x) = 0 & \text{for } (t, x) \in \sigma, \quad n = 1, 2, \dots \quad \text{and} \quad i \in S \end{cases} \tag{10.15}$$

and

$$\begin{cases} \mathcal{F}_c^i[v_n^i](t, x) = f^i(t, x, v_{n-1}(t, x), v_{n-1}) := \mathbf{F}^i[v_{n-1}](t, x), \quad i \in S, \quad \text{for } (t, x) \in D, \\ v_n^i(0, x) = 0 & \text{for } x \in G, \\ v_n^i(t, x) = 0 & \text{for } (t, x) \in \sigma, \quad n = 1, 2, \dots \quad \text{and} \quad i \in S. \end{cases} \tag{10.16}$$

From the form of infinite system (10.15) (and (10.16)), precisely from the diagonal form of differential operators $\mathcal{F}_c^i$, $i \in S$, it follows that this weakly coupled system has the following fundamental property: the ith equation depends on the ith unknown function only. Roughly speaking, system (10.15) (and (10.16)) is a collection of several equations each of which is an equation in one unknown function only. Therefore, its solution is a collection of separate solutions of the scalar problems considered in suitably selected space.

10.5 Applications

Let us consider the Fourier first initial-boundary value problem for an infinite system of equations

$$\begin{cases} \mathcal{F}_c^i[z^i](t, x) = g^i(t, x) & \text{for } (t, x) \in D, \\ z^i(0, x) = \phi_0^i(x) & \text{for } x \in G, \\ z^i(t, x) = h^i(t, x) & \text{for } (t, x) \in \sigma \quad \text{and} \quad i \in S \end{cases} \tag{10.17}$$

with the compatibility condition

$$h(0, x) = \phi_0(x) \quad \text{for } x \in \partial G,$$

where $z = \{z^i\}_{i \in S}$, $g = \{g^i\}_{i \in S}$, $\phi_0 = \{\phi_0^i\}_{i \in S}$, $h = \{h^i\}_{i \in S}$ and the differential operators $\mathcal{F}_c^i$, $i \in S$, are diagonal and uniformly parabolic in $\overline{D}$ and S is an arbitrary infinite (countable or uncountable) set of indices.

Using the above observations we will prove the following theorem:

Theorem 10.1 *On the existence and uniqueness of the solution of infinite systems of equations.*

Let us consider initial-boundary value problem (10.17) and assume that

i. *all the coefficients a_{kl}^i, b_k^i, c^i, $k, l = 1, \ldots, m$, $i \in S$ of the operators $\mathcal{L}_c^i$, $i \in S$, fulfill assumption $(H)^3$;*

ii. *the function g^i, $i \in S$, is uniformly Hölder continuous with respect to t and x in $\overline{D}$, i.e., $g \in C_S^{0+\alpha}(\overline{D})$;*

iii. *$\phi_0 \in C_S^{2+\alpha}(G)$, $h \in C_S^{2+\alpha}(\sigma)$, and $\partial G \in C^{2+\alpha}$.*

Then problem (10.17) has the unique solution u and $z \in C_S^{2+\alpha}(\overline{D})$.

Moreover, there exists a constant $C > 0$ depending only on the constants μ, K_1, α, and on the geometry of the domain D, such that the following a priori *Schauder estimate of the $(2 + \alpha)$-type holds*

$$\|z\|_{2+\alpha} \leq C \left(\|g\|_{0+\alpha} + \|\phi_0\|_{2+\alpha}^G + \|h\|_{2+\alpha}^\sigma \right). \tag{10.18}$$

Proof. Observe that system (10.17) has the following property: in the ith equation, only one unknown function with index i appears. Therefore, system (10.17) is a collection of individual independent equations. Applying known theorems from the literature on the existence and uniqueness of solutions of the Fourier first initial-boundary value problems for infinite systems of equations in Hölder spaces, we immediately obtain the estimates

$$|z^i|_{2+\alpha} \leq C_i \left(|g^i|_{0+\alpha} + \left|\phi_0^i\right|_{2+\alpha}^G + |h^i|_{2+\alpha}^\sigma \right) \quad \text{for } i \in S,$$

in which constants $C_i > 0$ are independent of g^i, ϕ_0^i, and h^i.

It follows directly that constants C_i depend on the constants μ, K_1, α only and on the geometry of the domain D. Constants C_i are also uniformly bounded for all $i \in S$. Therefore, there exists a constant $C > 0$ (C is independent of both the index i and the functions g^i, ϕ_0^i, and h^i) such that $C_i \leq C$, for all $i \in S$. Hence, by the definitions of the norms in the spaces $C_S^{0+\alpha}(\overline{D})$ and $C_S^{2+\alpha}(\overline{D})$, we obtain estimate (10.18). This completes the proof. $\square$

10.6 Conclusions

Let us emphasize that if we have one of the following boundary value problems for infinite (countable or uncountable) systems of equations with diagonal-diffusion nonlinear or

---

³ **Assumption (H).** We will assume that all the coefficients $a_{kl}^i = a_{kl}^i(t, x)$, $a_{kl}^i = a_{lk}^i$, $b_k^i = b_k^i(t, x)$, and $c^i = c^i(t, x)$ $(k, l = 1, \ldots, m, i \in S)$ of the operators $\mathcal{L}_c^i$ are uniformly Hölder continuous with respect to t and x in $\overline{D}$ with exponent α $(0 < \alpha < 1)$ and their Hölder norms are uniformly bounded, i.e.,
$$|a_{kl}^i|_{0+\alpha} \leq K_1, \quad |b_k^i|_{0+\alpha} \leq K_1, \quad |c^i|_{0+\alpha} \leq K_1.$$

differential operators with classical initial and boundary conditions, then we are able to consider this system as a collection of several equations each of which is an equation in one unknown only.

This is the main reason why we can apply standard topological tools to solve it. This observation is essential because it enables numerical computation to be simplified significantly.

Acknowledgment

We acknowledge the kind permission of Elsevier to allow us to reproduce in its entirety the following material: Brzychczy, S. and Górniewicz, L. (2011). Continuous and Discrete Models of Neural Systems in Infinite-Dimensional Abstract Spaces. *Neurocomputing*, Volume 74 (2011), pp. 2711–2715.

CHAPTER 11

Nonlinear Cable Equations

11.1 Introduction

In computational neuroscience it is in principle assumed that compartmentalization of neuronal structure leaves the dynamics exhibited by neurons and neural networks unchanged. This has led to a plethora of spiking neuron models which consist of isopotential compartments each containing a number of ionic channels modeled using the Hodgkin-Huxley formalism. This is known as the "lumped parameter assumption" and, as a result of this assumption, the spaced-clamped version of the Hodgkin-Huxley equations is taken to be an accurate representation of the entire neuron which is the basis of essentially all simulators that take neural morphology explicitly into account, like NEURON or GENESIS. The "lumped parameter assumption" suggests that if a continuously distributed system is divided into sufficiently small segments (or compartments) then one makes a negligible small error by assuming that each compartment is isopotential and uniform in its properties. The compartmental approach (the name being based on the compartmentalization of the neural process) was first developed in the 1960s in which the continuous partial differential equations of the continuous cable representation are replaced with a set of ordinary differential equations.

Spiking neurons as accurate exemplars of the dynamics of the real neurons employ discretization techniques first and then reduction techniques in justifying the reduced or spiking model captures the dynamics of the morphologically realistic neuron, but no comparison is made with the continuous model. One-dimensional continuum models provide a reference against which the accuracy of compartmental models may be assessed, and accurately define the misconceptions about the compartmental model and how it should be interpreted. In the finite difference approximation of the continuum model it is the result of numerical inaccuracy due to truncation error, while for a compartmental model it is due to the faithfulness of the "lumped parameter assumption." Like the cable model, the compartmental model is also exact in the sense that its formulation does not explicitly contain a procedure for specifying a level of approximation.

Mathematical Neuroscience. http://dx.doi.org/10.1016/B978-0-12-411468-5.00011-9

Infinitely extended dynamical systems with continuous space are represented in terms of the cable equations for each dendritic segment of the neuron's complex arbor. Infinitely extended dynamical systems with discrete space are lattice dynamical systems represented by finite-difference approximation to the continuous cable equation. Finitely extended dynamical systems are represented in terms of the spatially discrete analog of the continuous cable equation and reflect an extension of the nonextended (point) finite-dimensional dynamical systems (see Figure 11.1).

The solutions of infinite countable and infinite uncountable systems, which are discrete and continuous models of the same process, may be either identical or different, but nevertheless the truncation method indicated that the finite-difference approximations are close to the continuous model (see Chapter 6). In other words, use of finite-differences to approximate the cable equation can be shown by using the truncation method to bring about solutions that are reasonably close to each other. Nonetheless, the truncation method cannot confer whether the compartment model and the spiking neuron model, which are both spatially discrete analogs of the continuous cable model, yield solutions that are reasonably close to the continuous cable model.

The present chapter will focus on comparison theorems and maximum principles for the spatially discrete approximation of nonlinear cable models and applying the results to showing how such nonlinear cable models exhibit different dynamics to the spatially discrete analog of the continuous cable equation. Both spatially discrete models will be analyzed to describe the validity of the "lumped parameter assumption" used in compartmental models. The nonlinear analysis will lay the foundation for proving that compartmental models are poor exemplars of real neurons.

11.2 Nonlinear Cable Equations

The continuous cable equation may be expressed as

$$\frac{\partial V}{\partial t} = D\frac{\partial^2 V}{\partial x^2} + f(V, t), \tag{11.1}$$

where $V = V(x, t)$ is the membrane potential, x is the space along the cable (cm), t is time (ms), $D = \frac{\lambda^2}{\tau} = \frac{d}{4\rho C_m}$ is the diffusion coefficient in units of cm^2/ms, and f is a continuous function representing the reaction term, depending on both membrane potential and time.

The bistable equation is a special version of Eq. (11.1) when the functional is the cubic polynomial $f(V) := V(V - 1)(\tilde{a} - V)$, $0 < \tilde{a} < 1$, which arises in the Hodgkin-Huxley equations when the recovery variable n is held fixed at its steady-state, and the Hodgkin-Huxley model is bistable where $\tilde{a}$ is the bistability parameter. Cable equation (11.1) with the functional represented by a cubic polynomial is referred to as reduced FitzHugh-Nagumo equations or simply as Nagumo's equation.

The Nagumo equation has long been known to admit traveling waves. This equation still has traveling wave solutions, but they are traveling fronts instead of pulses. New analytical traveling wavefront solutions to the Nagumo's equation appeared. They were constructed with the use of different representations of three-zero kinetic function including the piecewise linear and sinusoidal approximations.

The discrete cable equation under the "lumped parameter assumption" corresponding to a compartmental model can be shown to be expressed as

$$\frac{dV_n}{dt} = d^*(V_{n+1} - 2V_n + V_{n-1}) + f(V_n), \tag{11.2}$$

where $d^* = \frac{D}{h^2}$ is the coupling coefficient with h as the distance between nodes and N is the total number of compartments. The study of the discrete cable equation (11.2) is substantially more difficult than that of the continuous cable equation (11.1). In the continuous cable equation, continuous changes in the physical parameters lead to continuous changes in the speed of propagation and the speed cannot be driven to zero, unless the diameter is zero. The propagation of the spatially discrete analog of the continuous equation may fail if the coupling coefficient is too small. Yet the spatially discrete analog of the continuous cable equation expresses traveling wave solutions when d^* is sufficiently large with dynamics similar to the continuous Nagumo equation.

However, there is a misconception that the "lumped parameter assumption" applies readily to the linear cable equation. When $f = 0$ the speed of the continuous problem (11.1) with $D = 1$ is $\frac{C_0}{h}$ where C_0 is the speed of propagation in the discretized problem (11.2).

11.3 Comparison of Solutions for Continuous and Discrete Cable Equations

In the 1970s there was significant progress made in using comparison and maximum principles to rigorously analyze the existence, uniqueness, and stability properties of traveling wave solutions to cable equations.

Comparison theorems play a very important role as differential inequalities techniques for cable equations and for their finite and infinite systems. These methods are commonly used in numerical methods for the computation of solutions and to prove the continuous dependence of the solutions upon the right-hand sides of the system and upon the initial and boundary conditions. Comparison theorems have also been used to prove the existence and uniqueness of solutions of cable equations with certain initial and boundary conditions.

The continuous Nagumo equation is used as a model for the propagation of nerve impulses or spikes without recovery. The propagating spike corresponds to a moving front or traveling wavefront and is sometimes referred to as a leading edge wavefront approximation to the nerve impulse.

Consider the continuous Nagumo equation for a semi-infinite cable:

$$U_t = DU_{xx} + f(U) \tag{11.3}$$

with the initial condition

$$U(0, x) = 0 \quad \text{for} \quad 0 \le x < \infty, \tag{11.4}$$

and boundary conditions

$$U(t, 0) = 1 \quad \text{for} \quad 0 \le t, \tag{11.5}$$

where the bistable source term $f = f(U)$ is a prescribed Lipschitz continuous function which is only a function of voltage (U) and not time. In particular, it can be given as the cubic polynomial

$$f(U) = U(U - 1)(\tilde{a} - U) \quad \text{for} \quad 0 \le \tilde{a} < 1. \tag{11.6}$$

Consider also the spatially discrete Nagumo equation formulated by the following differential-difference equation:

$$\frac{dV_n}{dt} = d^*(V_{n+1} - 2V_n + V_{n-1}) + f(V_n) \quad \text{for} \quad 2 \le n \le N - 1, \tag{11.7}$$

and with the initial-boundary conditions

$$V_1 = 1 \quad \text{for} \quad t \ge 0, \tag{11.8}$$

$$V_n = 0 \quad \text{for} \quad 2 \le n \le N - 1 \quad \text{and} \quad t = 0, \tag{11.9}$$

where $V_n \equiv V(t, nh)$. The discrete Nagumo equation depends on the coupling constant $d^* = \frac{D}{h^2}$ where h is the distance between nodes and the bistability parameter $\tilde{a}$.

Theorem 11.1 (Comparison theorem)   *Assume that*

   i.   *the function $U = U(t, x)$ is a bounded solution $U \in C^{2+\alpha}(\overline{D})$ of original problem (11.3–11.5), where the Hölder space $C^{2+\alpha}(\overline{D})$ is a Banach space of functions that are continuous in $\overline{D}$ and $0 < \alpha < 1$.*
   ii. *the function $V_n = V(t, nh)$ is a bounded solution of the original problem (11.7–11.9),*
   iii. *the function $f = f(y)$ fulfills the Lipschitz condition*

$$\mathcal{L}: \quad |f(\bar{y}) - f(y)| \le L_f \|\bar{y} - y\| \quad \text{for} \quad L_f > 0,$$

   iv. $d^* \cdot h^2 = D.$

From the above assumptions it follows that if there exists a constant $K > 0$ such that

$$|V_n(t) - U(t, nh)| \le K \cdot h^{\alpha} \quad \text{for all } n \in \mathbb{N},$$

then $\|V - U\| \to 0$ when $h \to 0$.

Proof. Denote $U_n(t) = U(t, nh)$. From (11.3) and (11.7) we get the error equation

$$\frac{d(V_n - U_n)}{dt} = d^*(V_{n+1} - 2V_n + V_{n-1}) + f(V_n) - \left[U_{xx}(t, nh) - f(U_n)\right]$$

$$= \frac{(V_{n+1} - U_{n+1}) - 2(V_n - U_n) + (V_{n-1} - U_{n-1})}{h^2}$$

$$+ \frac{U_{n+1} - 2U_n + U_{n-1}}{h^2} - U_{xx}(t, nh) + f(V_n) - f(U_n).$$

Denote

$$\Delta := \frac{U_{n+1} - 2U_n + U_{n-1}}{h^2} - U_{xx}(t, nh).$$

Therefore, from the mean value theorem for integrals we obtain

$$\Delta = \frac{1}{h^2} \int_0^h U_x(t, nh + \theta)d\theta - \frac{1}{h^2} \int_0^h U_x(t, nh - \theta)d\theta - U_{xx}(t, nh)$$

$$= \frac{1}{h^2} \int_0^h \int_{-\theta}^\theta U_{xx}(t, nh + s)ds d\theta - U_{xx}(t, nh).$$

Because $\frac{1}{h^2} \int_0^h \int_{-\theta}^\theta 1 ds d\theta = 1$ and $U \in C^{2+\alpha}(\overline{D})$, then

$$|\Delta| \leq \frac{1}{h^2} \int_0^h \int_{-\theta}^\theta |U_{xx}(t, nh + s) - U_{xx}(t, nh)| ds d\theta$$

$$\leq \frac{2}{h^2} \int_0^h \int_{-\theta}^\theta L_f |s^\alpha| ds d\theta$$

$$= 2\frac{L_f}{h^2} \int_0^h \int_{-\theta}^\theta \frac{\theta^{1+\alpha}}{1+\alpha} ds d\theta$$

$$= \frac{2L_f}{h^2} \frac{h^{2+\alpha}}{(1+\alpha)(2+\alpha)} = const \cdot h^\alpha.$$

Therefore, if $h \to 0$, then $\Delta \to 0$.

Consider the comparison system of ordinary differential equations

$$\frac{dZ_n}{dt} = \frac{Z_{n+1} - 2Z_n + Z_{n-1}}{h^2} + L_f Z_n + Const \cdot h^\alpha \qquad (11.10)$$

with the initial condition

$$Z_n(0) := \|V(0) - U(0)\|$$

$$= \sup_l |V_l(0) - U_l(0)|. \qquad (11.11)$$

Then Z_n tends to 0 as $h \to 0$, and $V_n - U_n$ tends uniformly to 0, because we have the estimate

$$|V_n(t) - U_n(t)| \leq Z_n(t) \quad \text{for all } n \in \mathbb{N}, \quad t \geq 0.$$

This completes the proof. $\qquad\qquad\qquad\qquad\qquad\qquad\qquad\qquad\qquad\qquad$ $\square$

Remark 11.1   Since $Z_n(0) = \text{const}$, then the comparison problem (11.10), (11.11) has the form of the following scalar Cauchy problem:

$$\frac{dy}{dt} = L_f y + Kh^\alpha, \quad y(0) = \|V(0) - U(0)\|.$$

Therefore, the only solution of (11.10), (11.11) is given by

$$Z_n(t) = y(t) = \|V(0) - U(0)\| e^{L_f t} + \frac{Kh^\alpha}{L_f} \left[e^{L_f t} - 1 \right].$$

Remark 11.2   The Hölder condition for U_{xx} can be weakened to a local Hölder condition, but the estimates become more difficult in that case. The Lipschitz condition for f can be weakened to a Perron condition.

Corollary 11.1   Under the assumption of Theorem 11.1, if U_{xx} is Lipschitz continuous in x, then

$$|V_n(t) - U(t, nh)| \leq \text{const} \cdot h \to 0,$$

provided that $|V_n(0) - U(0, nh)| \leq \text{const} \cdot h \to 0$.

Corollary 11.2   Under the assumption of Theorem 11.1, if U_{xxxx} is bounded, then

$$|V_n(t) - U(t, nh)| \leq \text{const} \cdot h^2 \to 0,$$

provided that $|V_n(0) - U(0, nh)| \leq \text{const} \cdot h^2 \to 0$.

11.4 Application of Comparison Theorem

Application of the comparison theorem to the discrete approximation of the continuous Nagumo equation shows that the behavior of the solution of the discrete Nagumo equation can be quite different from the solution of the continuous Nagumo equation.

Consider the discrete and continuous equations with the cubic right-hand side $f(U) = U(U - 1)(\tilde{a} - U)$, where $0 < \tilde{a} < 1$. The most interesting case $0 \leq U_0(x) \leq 1$ has the following property:

$$0 \leq U(t, x) \leq 1 \quad \text{for all } t \geq 0 \text{ and all } x.$$

The same property has the discrete version. This can be achieved by the comparison method, because 0 is a lower solution, whereas the upper solution satisfies the ordinary differential equation

$$\frac{d}{dt} Z = f(Z), \quad Z(0) = 1.$$

We are able to get effective estimates of the derivatives $U_x, U_{xx}, U_{xxx},$ and U_{xxxx} (note that the same estimates are valid for their discrete counterparts).

Lemma 11.1 *If the initial function U_0 is bounded in $[0, 1]$ with bounded derivatives in x of order 1 and 2, then there are constants $K_0, K_1, K_2, K_{2+\alpha} > 0$ such that $\|U\| \le K_0$,*

$$\|U_x(t, \cdot)\| \le K_1 e^t, \quad \|U_{xx}(t, \cdot)\| \le K_2 e^{2t},$$

$$\|U_{xx}(t, \bar{x}) - U_{xx}(t, x)\| \le K_{2+\alpha} e^{(2+\alpha)t} |\bar{x} - x|^\alpha.$$

Moreover if U_0 has bounded derivatives in x of order 3 and 4 then there exist positive constants K_3, K_4 such that

$$\|U_{xxx}(t, \cdot)\| \le K_3 e^{3t}, \quad \|U_{xxxx}(t, \cdot)\| \le K_4 e^{4t}.$$

Proof. From the above observation, we have $\|U\| \le 1 =: K_0$. Denote $U^{(1)} = U_x, U^{(2)} = U_{xx}$. Then these functions satisfy the equations

$$U_t^{(1)} - U_{xx}^{(1)} = U^{(1)} f'(U), \quad U^{(1)}(0, x) = U_{0,x}(x),$$

$$U_t^{(2)} - U_{xx}^{(2)} = U^{(2)} f'(U) + \left[U^{(1)}\right]^2 f''(U), \quad U^{(2)}(0, x) = U_{0,xx}(x).$$

Observe that $|f'(U)| \le 1$ for both the cases (quadratic and cubic). Thus we consider two comparison problems

$$\frac{d}{dt} Z^{(1)} \ge Z^{(1)}, \quad Z^{(1)}(0) \ge \|U_{0,x}\|,$$

$$\frac{d}{dt} Z^{(2)} \ge Z^{(2)} + 2\|f''\| \left[Z^{(1)}\right]^2, \quad Z^{(2)}(0) \ge \|U_{0,xx}\|.$$

One can easily find solutions to these comparison problems of the form

$$Z^{(1)}(t) = K_1 e^t, \quad Z^{(2)}(t) = K_2 e^{2t}.$$

The last comparison problem for $K_{2+\alpha}$ can be derived in a similar way.

We prove the estimates of U_{xxx} and U_{xxxx}. Denote $U^{(3)} = U_{xxx}$ and $U^{(4)} = U_{xxxx}$ then we can easily derive the parabolic equations for these functions. We write the respective comparison problems:

$$\frac{d}{dt} Z^{(3)} \ge Z^{(3)} + 3Z^{(1)} Z^{(2)} \|f''\| + (Z^{(1)})^3 \|f'''\|,$$

$$Z^{(3)}(0) \ge \|U_{0,xxx}\|,$$

$$\frac{d}{dt} Z^{(4)} \ge Z^{(4)} + 4Z^{(1)} Z^{(3)} \|f''\| + 3(Z^{(2)})^2 \|f''\| + 6(Z^{(1)})^2 Z^{(2)} \|f'''\| + (Z^{(1)})^4 \|f''''\|,$$

$$Z^{(4)}(0) \ge \|U_{0,xxxx}\|$$

from which we can easily find K_3 and K_4. This completes the proof. $\qquad\square$

This lemma together with the comparison theorem provides the following effective estimates:

Corollary 11.3 Consider the Nagumo equation and its discrete counterpart. If the initial function U_0 is bounded in $[0, 1]$ with bounded derivatives in x of order 1 and 2, then

$$|V_n(t) - U(t, nh)| \leq \frac{2K_{2+\alpha}}{(1+\alpha)(2+\alpha)} \cdot h^\alpha.$$

We have established the result that the solutions of (11.7) are very close to the respective solutions of (11.3). When the initial function is sufficiently regular, the right-hand side f is Lipschitz continuous and $d^* = 1/h^2$ tends to infinity. We claim that these properties are significantly different when some of these requirements are violated, e.g., the initial function has oscillations or it is unbounded. Consider (for simplicity) two examples of the case $f \equiv 0$.

Example   Fix $h > 0$. Consider the equation $U_t - U_{xx} = 0$ with h-periodic initial function U_0 such that $U_0(x) = 1$ on $[0, h/2)$ and $U_0(x) = 0$ on $[h/2, h)$. It is clear that the solution takes values in $[0, 1]$, and $\sup_x U(t, x) \to 1$ as $t \to \infty$. Consider now the respective discrete problem

$$\frac{d}{dt} V_n = \frac{V_{n+1} - 2V_n + V_{n-1}}{h^2}, \quad V_{2k}(0) = 1, \quad V_{2k+1}(0) = 0.$$

Since the solution is periodic, we reduce the problem to the system of two equations

$$\frac{d}{dt} V_0 = \frac{2V_1 - 2V_0}{h^2}, \quad \frac{d}{dt} V_1 = -\frac{2V_1 - 2V_0}{h^2}$$

with the initial values $V_0(0) = 1$, $V_1(0) = 0$. It is clear that the solution

$$V_0 = \frac{1 + e^{-4t/h^2}}{2}, \quad V_1 = \frac{1 + e^{-4t/h^2}}{2}$$

rapidly tends to $1/2$.

Example   In linear cable theory, it is assumed that for an initially quiescent cable $U(0, x) = 0$, but for a nonquiescent cable, the linear cable equation becomes a Cauchy problem:

$$U_t - U_{xx} = 0, \quad U(0, x) = e^{x^2}.$$

It is well known that this problem has the unique solution in the classes $|U(t, x)| \leq const \cdot e^{Kx^2}$, for any $K > 1$. We will show that the respective discrete problem has no solution in such natural classes, thus it is ill posed.

Theorem 11.2   *Let $h > 0$. The problem*

$$\frac{d}{dt} V_n = \frac{V_{n+1} - 2V_n + V_{n-1}}{h^2}, \quad V_n(0) = e^{h^2 n^2}$$

cannot have any solutions satisfying the growth condition $|V_n| \leq const \cdot e^{Kh^2 n^2}$, where $K > 1$.

Proof.   Suppose (on the contrary) that there exists such a solution, then it is nonnegative. We will show that this solution is arbitrarily large. It is convenient to introduce new dependent variables $W_n := V_n e^{2t/h^2}$. Then we have the system

$$\frac{d}{dt} W_n = \frac{W_{n+1} + W_{n-1}}{h^2}, \quad W_n(0) = e^{h^2 n^2}.$$

Each function W_n is increasing in t, because the right-hand side is positive. Hence

$$W_n \geq W_n(0) = e^{h^2 n^2}.$$

By symmetry $W_n = W_{-n}$ we get the inequality

$$W_n = W_n(0) + \int_0^t \frac{W_{n+1}(s) + W_{n-1}(s)}{h^2} ds \geq \int_0^t \frac{W_{|n|+1}(s)}{h^2} ds.$$

By induction on k we can prove the inequality

$$W_n \geq \frac{t^k}{k! \, h^{2k}} e^{h^2 (|n|+k)^2}.$$

The sequence tends to infinity, which completes the proof.   □

To summarize, if the initial function is bounded, together with first and second derivatives then $V_n \to U$. If the data is unbounded or highly irregular then $V_n \nrightarrow U$.

11.5 Conclusions

Cable properties of dendrites are ubiquitous to all biological neural networks. They allow for (i) the spatial distribution of synapses, (ii) the spatial distribution of ionic channels, (iii) second-messengers, and (iv) endogenous membranous structure. Yet there is a propensity to model spiking neurons with a single compartment. The definition of a network composed of a single cell compartment is a trait of the computational modelers who persist because they are incapable of incorporating cable properties of single neurons into realistic network models. When continuous cable properties are filtered by way of compartments or otherwise used in artificial neural networks, the unproven assumption is that dynamics remain the same. However, the transition gives a piecemeal description of the dynamics of neural networks in their natural environment. It foreshadows the differences between finite- and infinite-dimensional nonlinear neurodynamical systems, as we have shown.

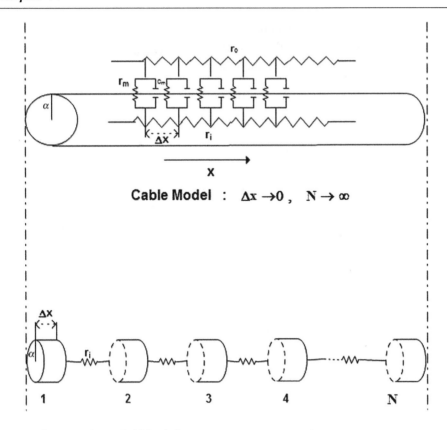

Cable Model : $\Delta x \to 0$, $N \to \infty$

Compartmental Model : $\Delta x \to 0$, N is finite

Figure 11.1 Spatially discrete approximation of the continuous cable (top) and spatially discrete analog of the continuous cable where the cable is viewed as *N* discrete entities or compartments (bottom).

Acknowledgment

We acknowledge the kind permission of World Scientific Publishing Co Pte Ltd to allow us to reproduce in its entirety the following material: Brzychczy, S., Leszcynski, H., and Poznanski, R.R. (2012). Neuronal Models in Infinite-Dimensional Spaces and their Finite-Dimensional Projections, Part 2. *Journal of Integrative Neuroscience*, Volume 11, pp. 265–276.

CHAPTER 12

Reaction-Diffusion Equations

12.1 Introduction

The classical view is that the solutions of reaction-diffusion equations display a wide range of behaviors of importance in neuroscience, including the formation of traveling waves in nerve conduction, i.e., traveling wave solutions of the full Hodgkin-Huxley equations. In order to study the Hodgkin-Huxley equations, a general theory of quasilinear reaction-diffusion equations is needed.

We consider reaction-diffusion equations of the following form:

$$\frac{\partial z}{\partial t} - Q(z)z = f\left(t, x, z, \frac{\partial z}{\partial x}\right), \tag{12.1}$$

where the quasilinear operator $Q(z)$ of the general form is defined by

$$Q(z)z := \sum_{j,k=1}^{m} a_{jk}\left(t, x, z, \frac{\partial z}{\partial x}\right) \frac{\partial^2 z}{\partial x_j \partial x_k} \tag{12.2}$$

and

$$\frac{\partial z}{\partial t} - \mathfrak{A}(z)z = f\left(t, x, z, \frac{\partial z}{\partial x}\right), \tag{12.3}$$

where the quasilinear operator $\mathfrak{A}(z)$ of the divergence form is defined by

$$\mathfrak{A}(z)z := \sum_{j=1}^{m} \frac{\partial}{\partial x_j} d_j\left(t, x, z, \frac{\partial z}{\partial x}\right), \tag{12.4}$$

and the coefficients of these operators are defined for all $(t, x, y, p) \in D \times \mathbb{R} \times \mathbb{R}^m$, where $D := (0, \infty) \times G$ and G is a bounded domain in $\mathbb{R}^m$ whose boundary ∂G is sufficiently smooth.

While studying quasilinear reaction-diffusion equations, from the very beginning a difficulty arises with the dependence of the coefficients in the quasilinear operators $Q(z)$ and $\mathfrak{A}(z)$ on the unknown solution z and its first spatial derivatives $\frac{\partial z}{\partial x}$. Thus the families of operators $Q(z)$

and $\mathfrak{A}(z)$ appearing in Eqs. (12.1) and (12.3) depend on the unknown solution z and this dependence renders the equations nonlinear with respect to z and $\frac{\partial z}{\partial x}$. Under such circumstances, already the definition of the domain of such an operator proves difficult. Two cases are possible: either the operator domain does not change with the function z or it does.

The behavior of solutions of quasilinear reaction-diffusion equations is substantially different from that of solutions of semilinear reaction-diffusion equations. Nonlinear problems reveal both certain qualitative features known in the theory of linear equations, and certain features definitely nonlinear in their nature, including: blowup, quenching, and dead core. In recent years quasilinear reaction-diffusion equations and finite systems thereof have been studied by numerous authors.

Throughout this chapter we confine ourselves to defining the ellipticity and parabolicity of quasilinear reaction-diffusion equations and presenting simple examples thereof.

12.2 Ellipticity and Parabolicity

Let us consider the reaction-diffusion equation arising frequently in neuroscience

$$\frac{\partial z}{\partial t} - \tilde{\mathfrak{A}}(z)z = f(t, x, z) \quad \text{for } (t, x) \in D := (0, +\infty) \times G, \quad z \in \mathbb{R}, \tag{12.5}$$

where the quasilinear operator $\tilde{\mathfrak{A}}(z)$ of the special form is defined by

$$\tilde{\mathfrak{A}}(z)z := \sum_{j,k=1}^{m} \frac{\partial}{\partial x_j}\left(d_{jk}(t, x, z)\frac{\partial z}{\partial x_k}\right). \tag{12.6}$$

If we assume that the coefficients $d_{jk} = d_{jk}(t, x, z)$, $d_{jk} = d_{kj}$, $j, k = 1, 2, \ldots, m$, are C^1-continuous with respect to x and z for $(t, x, z) \in D \times \mathbb{R}$ and $z \in C^1(D)$, then we will write Eq. (12.5) in the following equivalent form:

$$\frac{\partial z}{\partial t} - A(z)z = f(t, x, z) \quad \text{for} \quad (t, x, z) \in D \times \mathbb{R}, \tag{12.7}$$

where

$$\begin{aligned}
\tilde{\mathfrak{A}}(z)z &:= \sum_{j,k=1}^{m} \frac{\partial}{\partial x_j}\left(d_{jk}(t, x, z)\frac{\partial z}{\partial x_k}\right) \\
&= \sum_{j,k=1}^{m}\left[\left(\frac{\partial d_{jk}}{\partial x_j} + \frac{\partial d_{jk}}{\partial z}\frac{\partial z}{\partial x_j}\right)\frac{\partial z}{\partial x_k} + d_{jk}\frac{\partial^2 z}{\partial x_j \partial x_k}\right]
\end{aligned}$$

$$= \sum_{k=1}^{m} \left(\sum_{j=1}^{m} \frac{\partial d_{jk}}{\partial x_j} \right) \frac{\partial z}{\partial x_k} + \sum_{j,k=1}^{m} \frac{\partial d_{jk}}{\partial z} \frac{\partial z}{\partial x_j} \frac{\partial z}{\partial x_k} + \sum_{j,k=1}^{m} d_{jk} \frac{\partial^2 z}{\partial x_j \partial x_k} \qquad (12.8)$$

$$= \sum_{j,k=1}^{m} d_{jk}(t, x, z) \frac{\partial^2 z}{\partial x_j \partial x_k} + \sum_{j,k=1}^{m} c_{jk}(t, x, z) \frac{\partial z}{\partial x_j} \frac{\partial z}{\partial x_k} + \sum_{k=1}^{m} d_k(t, x, z) \frac{\partial z}{\partial x_k}$$

$$:= \tilde{A}(z)z.$$

On the other hand, if the operator $\tilde{A}(z)$ is defined by

$$\tilde{A}(z)z := \sum_{j,k=1}^{m} a_{jk}(t, x, z) \frac{\partial^2 z}{\partial x_j \partial x_k} + \sum_{j,k=1}^{m} b_{jk}(t, x, z) \frac{\partial z}{\partial x_j} \frac{\partial z}{\partial x_k} + \sum_{j=1}^{m} c_j(t, x, z) \frac{\partial z}{\partial x_j}, \quad (12.9)$$

and the coefficients in the principal part of operator $\tilde{\mathfrak{A}}$ are sufficiently smooth then this operator can be transformed to the equivalent form $\tilde{\mathfrak{A}}(z)$ given by (12.6).

It is well known that in the case of linear operators, the transition from the general form of a second-order differential operator to the divergence form thereof, as well as the reverse transition are feasible, provided that the coefficients in the principal part of relevant operators are sufficiently smooth, under which condition the forms are equivalent.

But the quasilinear reaction-diffusion equations with principal part in the divergence form (12.4) do not include an entire class of quasilinear reaction-diffusion equations in the general form given by (12.2), because not every quasilinear operator in the general form with smooth coefficients can be transformed into an operator in the divergence form.

Now, analogously, we adopt the following definitions of the ellipticity and parabolicity of quasilinear reaction-diffusion equations of the second order.

Let us consider quasilinear reaction-diffusion equation (12.1), i.e.,

$$\frac{\partial z}{\partial t} - Q(z)z = f\left(t, x, z, \frac{\partial z}{\partial x}\right),$$

where the operator $Q(z)$ is defined by (12.2), i.e.,

$$Q(z)z := \sum_{j,k=1}^{m} a_{jk}\left(t, x, z, \frac{\partial z}{\partial x}\right) \frac{\partial^2 z}{\partial x_j \partial x_k},$$

and the coefficients $a_{jk}(t, x, y, p)$, $a_{jk} = a_{kj}$, $j, k = 1, 2, \ldots, m$ of $Q(z)$ are defined for $(t, x, y, p) \in D \times \mathbb{R} \times \mathbb{R}^m := \Omega$.

Let Λ be a subset of Ω. Then the quasilinear operator $Q(z)$ is called *elliptic* in Λ if and only if the coefficient matrix $(a_{jk}(t, x, y, p))$ is positive defined for all $(t, x, y, p) \in \Lambda$. This means that if $\mu(t, x, y, p)$ and $\nu(t, x, y, p)$ denote the minimum and maximum eigenvalues, respectively, of the coefficient matrix $(a_{jk}(t, x, y, p))$, then:

$$0 < \mu(t, x, y, p) |\xi|^2 \le \sum_{j,k=1}^{m} a_{jk}(t, x, y, p)\xi_j\xi_k \le \nu(t, x, y, p) |\xi|^2$$

for all $(t, x, y, p) \in \Lambda$ and for all $\xi \in \mathbb{R}^m \setminus \{0\}$, where $\xi = (\xi_1, \xi_2, \dots, \xi_m)$ (here and below). If ν/μ is bounded in $\overline{\Lambda}$ then we shall call $Q(z)$ uniform elliptic in $\overline{\Lambda}$.

The other condition of uniform ellipticity of the operator $Q(z)$ in $\overline{\Lambda}$ is of the following form:

$$0 \le \mu(|y|) |\xi|^2 \le \sum_{j,k=1}^{m} a_{jk}(t, x, y, p)\xi_j\xi_k \le \nu(|y|) |\xi|^2$$

for all $(t, x, y, p) \in \overline{\Lambda}$ and for all $\xi \in \mathbb{R}^m$, where $\mu(\tau)$ is a positive nonincreasing continuous function defined for $\tau \ge 0$ and $\nu(\tau)$ is an arbitrary nondecreasing continuous function defined for $\tau \ge 0$.

If the operator $Q(z)$ is elliptic (uniformly elliptic) in Λ, then the operator $\mathcal{F}(z) = \frac{\partial}{\partial t} - Q(z)$ is called parabolic (uniformly parabolic) in Λ and the quasilinear reaction-diffusion equation (12.1) is called parabolic in Λ.

Let us consider the quasilinear reaction-diffusion equation (12.1), where the operator $Q(z)$ is defined by (12.2). If the function $u = u(t, x) \in C^1(D)$ and the coefficient matrix $\left(a_{jk}^i\left(t, x, u(t, x), \frac{\partial u(t,x)}{\partial x}\right)\right)$ is positive defined for all $(t, x) \in D$, then we say that the operator $Q(u)$ *is elliptic with respect to the function u in D.*

For the quasilinear reaction-diffusion equation (12.3) with the principal part in divergence form (12.4) the condition of uniform ellipticity of the operator $\mathfrak{A}(z)$ in Ω is of the following form:

$$\mu(|y|)\left(1 + |p|^{m-2}\right) |\xi|^2 \le \sum_{j,k=1}^{m} \frac{\partial d_j(t, x, y, p)}{\partial p_k}\xi_j\xi_k \le \nu(|y|)\left(1 + |p|^{m-2}\right) |\xi|^2$$

for all $(t, x, y, p) \in \Omega$ and for all $\xi \in \mathbb{R}^m$, where m is a number greater than 1 and $p = (p_1, p_2, \dots, p_m)$ and functions $\mu(\tau)$ and $\nu(\tau)$ are defined as above.

12.3 Transformation of Reaction-Diffusion Equations

Let us consider the linear diffusion equation in homogeneous isotropic medium[1]

$$c\rho\frac{\partial z}{\partial t} - \nabla \cdot (\kappa \nabla z) := c\rho\frac{\partial z}{\partial t} - \sum_{j=1}^{3}\frac{\partial}{\partial x_j}\left(\kappa\frac{\partial z}{\partial x_j}\right) = f(t, x, z), \tag{12.10}$$

where $x = (x_1, x_2, x_3) \in G \subset \mathbb{R}^3$, ∇ is the gradient operator in $\mathbb{R}^3$, i.e., $\nabla = \left(\frac{\partial}{\partial x_1}, \frac{\partial}{\partial x_2}, \frac{\partial}{\partial x_3}\right)$, G is a bounded domain representing the conduction medium (material), the unknown function $z = z(t, x)$, and f is the reaction function (the internal sources).

The dependence $\kappa = \kappa(z)$ may not be omitted without a material effect on the solution. The situation is more complicated, since the equation becomes a quasilinear reaction-diffusion equation.

In fact, if we assume that $\kappa = \kappa(z)$ is a C^1-function of z for $z \geq 0$ and $\kappa(z) \geq \delta_0$ for some constant $\delta_0 > 0$ then (12.10) becomes the following quasilinear reaction-diffusion equation in isotropic medium

$$c\rho\frac{\partial z}{\partial t} - \kappa(z)\sum_{j=1}^{3}\frac{\partial^2 z}{\partial x_j^2} - \frac{d\kappa}{dz}\sum_{j=1}^{3}\left(\frac{\partial z}{\partial x_j}\right)^2 = f(t, x, z) \tag{12.11}$$

for $(t, x, y) \in (0, T] \times G \times \mathbb{R}$.

In the case of *anisotropic medium* $\kappa_{jk} = \kappa_{jk}(z)$, we obtain the *quasilinear reaction-diffusion equation* (with cubic anisotropy)

$$c\rho\frac{\partial z}{\partial t} - \sum_{j,k=1}^{3}\kappa_{jk}(z)\frac{\partial^2 z}{\partial x_j \partial x_k} - \sum_{j,k=1}^{3}\frac{d\kappa_{jk}}{dz}\left(\frac{\partial z}{\partial x_j}\right)\left(\frac{\partial z}{\partial x_k}\right) = f(t, x, z). \tag{12.12}$$

Let us consider this equation in dimension one, i.e., the equation

$$c\rho\frac{\partial z}{\partial t} - \kappa(z)\frac{\partial^2 z}{\partial x^2} - \frac{d\kappa}{dz}\left(\frac{\partial z}{\partial x}\right)^2 = f(t, x, z) \quad \text{for } t > 0, \quad x \in \mathbb{R}. \tag{12.13}$$

Then, under appropriate assumptions, quasilinear reaction-diffusion equation (12.13) may be transformed into a semilinear reaction-diffusion equation using the transformation

$$w = \Phi(z) := \int_{z_0}^{z}\frac{\kappa(s)}{\kappa_0}ds, \tag{12.14}$$

---

[1] If $a = (a_1, a_2, \ldots, a_m)$, $b = (b_1, b_2, \ldots, b_m) \in \mathbb{R}^m$, then the inner product is defined as follows: $a \cdot b = \sum_{j=1}^{m} a_j b_j$.

where z_0 is a nonnegative constant and $\kappa_0 = \kappa(z_0)$. Then

$$\frac{dw}{dz} = \frac{\kappa(z)}{\kappa_0}, \quad \frac{\partial w}{\partial t} = \frac{\kappa(z)}{\kappa_0}\frac{\partial z}{\partial t}, \quad \frac{\partial w}{\partial x} = \frac{\kappa(z)}{\kappa_0}\frac{\partial z}{\partial x}, \tag{12.15}$$

$$\frac{\partial^2 w}{\partial x^2} = \frac{1}{\kappa_0}\frac{d\kappa}{dz}\left(\frac{\partial z}{\partial x}\right)^2 + \frac{\kappa(z)}{\kappa_0}\frac{\partial^2 z}{\partial x^2}$$

and from Eq. (12.13) we obtain the following semilinear reaction-diffusion equation:

$$\frac{\partial w}{\partial t} - \frac{\kappa_0}{c\rho}\frac{\partial^2 w}{\partial x^2} = \frac{\kappa_0}{c\rho}f(t, x, z). \tag{12.16}$$

Because

$$\frac{dw}{dz} = \frac{d\Phi}{dz} = \frac{\kappa(z)}{\kappa_0} > 0$$

then there exists the function $z = \Phi^{-1}(w) = \eta(w)$ inverse to the function $w = \Phi(z)$ and it is an increasing function of $w > 0$. Therefore, Eq. (12.13) is equivalent to a coupled system of Eq. (12.16) and the algebraic equation $z = \eta(w)$

$$\begin{cases} \dfrac{\partial w}{\partial t} - \dfrac{\kappa_0}{c\rho}\dfrac{\partial^2 w}{\partial x^2} = \dfrac{\kappa_0}{c\rho}f(t, x, z), \\ z = \eta(w). \end{cases} \tag{12.17}$$

It is clear that a pair (z, w) is a solution of system (12.17) (with suitable initial-boundary conditions) in $\overline{D}$ if and only if z is a solution of quasilinear reaction-diffusion equation (12.13) (with the same conditions) in $\overline{D}$.

Next, if we additionally assume that $c = c(z)$ and $c(z) > 0$ for $z \geq 0$, then we introduce the function H (enthalpy), setting

$$v = H(z) := \int_{z_0}^{z} c(s)ds. \tag{12.18}$$

There is

$$\frac{\partial v}{\partial t} = c(z)\frac{\kappa_0}{\kappa(z)}\frac{\partial w}{\partial t}$$

and finally we obtain the semilinear parabolic equation of the form

$$\frac{\partial v}{\partial t} - \frac{\kappa_0}{\rho}\frac{\partial^2 w}{\partial x^2} = \frac{\kappa_0}{c_0}f(t, x, z), \tag{12.19}$$

where $c_0 = c(z_0)$. Quasilinear reaction-diffusion equation (12.13) is equivalent to a coupled system of equation (12.19) and two algebraic equations

$$
\begin{cases}
\dfrac{\partial v}{\partial t} - \dfrac{\kappa_0}{\rho} \dfrac{\partial^2 w}{\partial x^2} = \dfrac{\kappa_0}{c \rho_0} f(t, x, z), \\
z = \eta(w), \\
v = H(z).
\end{cases}
\tag{12.20}
$$

The two auxiliary functions $w = \Phi(z)$ and $v = H(z)$ defined by (12.14) and (12.18) have been introduced to transform quasilinear equation (12.13) into a simpler form of semilinear reaction-diffusion equation (12.16), and subsequently (12.19), which (with initial and boundary conditions included) can easily be solved with a monotone iterative method.

Let us note that the method described above and the results obtained may serve as a starting point for the investigation into certain classes of quasilinear reaction-diffusion equations.

A class of quasilinear reaction-diffusion equations in the form

$$
q(t, x, z) \frac{\partial z}{\partial t} - \nabla \cdot \big(\kappa(t, x, z) \nabla z\big) = f_0(t, x, z)
\tag{12.21}
$$

in a cylindrical domain $D := (0, T] \times G$, where T is an arbitrary positive constant and G is a bounded domain in $\mathbb{R}^m$ representing the conduction medium, whose boundary is of class $C^{1+\alpha}$ $(0 < \alpha < 1)$, $z = z(t, x)$ is the concentration of ionic species at the point $(t, x) \in \overline{D} := [0, T] \times \overline{G}$, with the initial condition

$$
z(0, x) = \phi_0(x) \quad \text{in } G,
\tag{12.22}
$$

and the Dirichlet boundary condition

$$
z(t, x) = g_0(t, x) \quad \text{on } \sigma,
\tag{12.23}
$$

or the nonlinear Neumann boundary condition

$$
\kappa(t, x, z) \frac{dz}{dv} = g_0(t, x, z) \quad \text{on } \sigma,
\tag{12.24}
$$

where the initial and boundary conditions satisfy a suitable compatibility zero-order condition.

To solve this problem, a monotone iterative method, as in the case of semilinear reaction-diffusion equations is used. To this end, additional assumptions are necessary: that the right-hand sides of (12.21) and (12.24) are monotone with respect to z, as well as assumption A_0.[2]

---

[2] Assumption A_0 implies the existence of at least one pair of sufficiently smooth functions u_0, v_0 in $\overline{D}$, which are lower and upper solutions, respectively, of problem (12.21)–(12.23) in $\overline{D}$ and $u_0 \leq v_0$. The set $\langle u_0, v_0 \rangle$ is a function interval (sector) defined by the functions u_0 and v_0 in a suitable function space.

Assuming the function $\kappa = \kappa(t, x, z)$ to be of class C^1 with respect to x and z in $D \times \langle u_0, v_0 \rangle$ we have

$$
\begin{aligned}
\nabla \cdot \left[\kappa(t, x, z) \nabla z \right] &= \frac{\partial}{\partial x_1} \left(\kappa(t, x, z) \frac{\partial z}{\partial x_1} \right) + \frac{\partial}{\partial x_2} \left(\kappa(t, x, z) \frac{\partial z}{\partial x_2} \right) \\
&\quad + \cdots + \frac{\partial}{\partial x_m} \left(\kappa(t, x, z) \frac{\partial z}{\partial x_m} \right)
\end{aligned}
\tag{12.25}
$$

$$
= \kappa(t, x, z) \sum_{j=1}^{m} \frac{\partial^2 z}{\partial x_j^2} + \frac{\partial \kappa}{\partial z} \sum_{j=1}^{m} \left(\frac{\partial z}{\partial x_j} \right)^2 + \sum_{j=1}^{m} \frac{\partial \kappa}{\partial x_j} \frac{\partial z}{\partial x_j}
$$

$$
:= \mathbf{A}(z)z.
\tag{12.26}
$$

Using earlier computations, we arrive at the equation

$$
q(t, x, z) \frac{\partial z}{\partial t} - \mathbf{A}(z)z = f_0(t, x, z) \quad \text{for } (t, x, z) \in D \times \langle u_0, v_0 \rangle,
\tag{12.27}
$$

where the operator $\mathbf{A}(z)$ is defined by (12.26).

The key feature of the monotone iterative method is that a solution of a nonlinear problem is defined as the limit of monotone sequences of solutions of linear problems, with the iterative process starting from the initial iterations u_0 and v_0. Unfortunately, in (12.27), there appears the nonlinear term $\sum_{j=1}^{m} \left(\frac{\partial z}{\partial x_j} \right)^2$, which means that the corresponding iterative process is nonlinear.

To negotiate this difficulty, we confine the study to a class of quasilinear reaction-diffusion equations received under the fundamental assumption there exists a positive function $p = p(z)$ defined in the sector $\langle u_0, v_0 \rangle$ for $z \neq 0$ and positive functions $q_0 = q_0(t, x)$ and $\kappa_0 = \kappa_0(t, x)$ defined in D are given as products:

$$
\begin{aligned}
q(t, x, z) &= q_0(t, x) p(z), \\
\kappa(t, x, z) &= \kappa_0(t, x) p(z) \quad \text{for } (t, x) \in D \quad \text{and} \quad z \in \langle u_0, v_0 \rangle.
\end{aligned}
\tag{12.28}
$$

Having adopted this extremely restrictive assumption, problems (12.21), (12.24), and (12.22) are reduced to the problem

$$
\begin{cases}
p(z) \dfrac{\partial z}{\partial t} - \dfrac{1}{q_0(t, x)} \nabla \cdot \left(\kappa_0(t, x) p(z) \nabla z \right) = f(t, x, z) \quad \text{for } (t, x) \in D, \\[2mm]
p(z) \dfrac{dz}{dt} = g(t, x, z) \quad \text{for } (t, x) \in \sigma, \\[2mm]
z(0, x) = z_0(x) \quad \text{for } x \in G,
\end{cases}
\tag{12.29}
$$

where

$$f(t, x, z) := \frac{1}{q_0(t, x)} f_0(t, x, z),$$

$$g(t, x, z) := \frac{1}{q_0(t, x)} g_0(t, x, z) \quad \text{for } (t, x) \in D \quad \text{and} \quad z \in \langle u_0, v_0 \rangle .$$

Because

$$\frac{1}{q_0(t, x)} \nabla \cdot \left(\kappa_0(t, x) p(z) \nabla z \right) = \frac{1}{q_0(t, x)} \left[\nabla \kappa_0 \cdot (p(z) \nabla z) + \kappa_0(t, x) \nabla \cdot (p(z) \nabla z) \right]$$

$$= a \nabla \cdot \left(p(z) \nabla z \right) + \vec{c} \cdot \left(p(z) \nabla z \right),$$

where $a = \dfrac{\kappa_0(t, x)}{q_0(t, x)}$, $\vec{c} = \dfrac{\nabla \kappa_0(t, x)}{q_0(t, x)}$, then the problem (12.29) has the form

$$\begin{cases} p(z) \dfrac{\partial z}{\partial t} - \left[\left(a \nabla \cdot (p(z) \nabla z) \right) + \vec{c} \cdot (p(z) \nabla z) \right] = f(t, x, z) & \text{for } (t, x) \in D, \\ p(z) \dfrac{dz}{dt} = g(t, x, z) & \text{for } (t, x) \in \sigma, \\ z(0, x) = z_0(x) & \text{for } x \in G, \end{cases} \tag{12.30}$$

If we define

$$w = \Phi(z) := \int_0^z p(s) ds \quad \text{for } z \in \langle u_0, v_0 \rangle \tag{12.31}$$

then $\frac{dw}{dz} = p(z) > 0$. Therefore, there exists the function $z = \Phi^{-1}(w) := \Psi(w)$ inverse to the function $w = \Phi(z)$ and is an increasing function on w.

There is

$$\frac{\partial w}{\partial t} = p(z) \frac{\partial z}{\partial t}, \quad \nabla w = p(z) \nabla z, \quad \frac{dw}{dv} = p(z) \frac{dz}{dv} \tag{12.32}$$

and we may write problem (12.29) as a coupled system of the semilinear reaction-diffusion equation with the initial and boundary conditions and an algebraic equation, in the form

$$\begin{cases} \dfrac{\partial w}{\partial t} - a \nabla^2 w - \vec{c} \cdot \nabla w = f(t, x, z) & \text{for } (t, x) \in D, \\ \dfrac{dw}{dv} = g(t, x, z) & \text{for } (t, x) \in \sigma, \\ w(0, x) = w_0(x) & \text{for } x \in G, \\ z = \Psi(w) & \text{for } (t, x) \in D, \end{cases} \tag{12.33}$$

where $z \in \langle u_0, v_0 \rangle$ and $w_0(x) = \Phi(z_0(x))$ for $x \in G$.

It is clear that (z, w) is a solution of problem (12.33) if and only if z is a solution of problems (12.21), (12.24), and (12.22) in $\overline{D}$, and problem (12.33) may be solved with no serious difficulties.

12.4 Reaction-Diffusion Equations in Diffusion Processes

The equation of the form

$$\frac{\partial z}{\partial t} - \sum_{j=1}^{3} \frac{\partial}{\partial x_j}\left(d\frac{\partial z}{\partial x_j}\right) = f(t, x, z), \tag{12.34}$$

where $x = (x_1, x_2, x_3)$, $z = z(t, x)$ is the density function, and d is the diffusion coefficient, is a *reaction-diffusion equation*.

In numerous reaction-diffusion problems, the reaction term f depends on the density function z. This is known as facilitated diffusion. It occurs when the flux of an ionic species is amplified by a reaction that takes place in the diffusing medium. If the diffusion coefficient d depends on the density, i.e., $d = d(z) > 0$ for $z \geq 0$, then we obtain the quasilinear reaction-diffusion equation of the form

$$\frac{\partial z}{\partial t} - \sum_{j=1}^{3} \frac{\partial}{\partial x_j}\left(d(z)\frac{\partial z}{\partial x_j}\right) = f(t, x, z). \tag{12.35}$$

If the diffusion process is *time-dependent* (or *non-stationary*) then the density function z depends on t, i.e., $z = z(t, x)$ and we consider diffusion equation (12.35) which is called the *time-dependent* (or *non-stationary*) *reaction-diffusion equation*. If the reaction process is *steady-state*, then the density function z is independent of t, i.e., $z = z(x)$ and we consider an equation of the form

$$-\sum_{j=1}^{3} \frac{\partial}{\partial x_j}\left(d(z)\frac{\partial z}{\partial x_j}\right) = f(x, z), \tag{12.36}$$

which is called the *steady-state* (or *stationary*) *reaction-diffusion* equation.[3]

Fundamental information on the comparison theorem techniques can be used to prove the existence and uniqueness of solutions of *multicomponent (precisely: two-component) diffusion systems* of the form

$$\begin{cases} \dfrac{\partial u}{\partial t} - \mathcal{L}_c[u](t, x) = f(t, x, u, w), \\[2mm] \dfrac{\partial w}{\partial t} = g(t, x, u, w), \end{cases} \tag{12.37}$$

where

$$\mathcal{L}_c := \sum_{j,k=1}^{m} a_{jk}(t, x)\frac{\partial^2}{\partial x_j \partial x_k} + \sum_{j=1}^{m} b_j(t, x)\frac{\partial}{\partial x_j} + c(t, x)\mathcal{I}$$

---

[3] This terminology also applies to equations describing other processes.

for $(t, x) \in D := (0, T] \times G$, where G is an open and bounded domain in $\mathbb{R}^m$. We assume the function f is a nondecreasing function with respect to w, but g is a nonincreasing function with respect to u and both satisfy the Lipschitz condition with respect to u and w, the coefficients of the uniformly elliptic operator $\mathcal{L}_c$ in $\overline{D}$ are sufficiently regular with the initial and boundary conditions of the Dirichlet type.

The comparison theorem for a particular class of multicomponent diffusion systems can be readily obtained as follows:

$$\frac{\partial u^i(t, x)}{\partial t} - \mathcal{L}_c^i[u^i](t, x) = f^i(t, x, u), := f^i(t, x, u^1, u^2, \ldots, u^r), \quad i = 1, 2, \ldots, r,$$

(12.38)

where the operators $\mathcal{L}_c^i$ have the form

$$\mathcal{L}_c^i := \sum_{j,k=1}^m a_{jk}^i(x) \frac{\partial^2}{\partial x_j \partial x_k} + \sum_{j=1}^m b_j^i(x) \frac{\partial}{\partial x_j} + c^i(x)\mathcal{I}$$

are uniformly elliptic in $\overline{D}$, with the initial and boundary conditions

$$z(0, x) = \phi_0(x) \quad \text{for } x \in G,$$
$$z(t, x) = 0 \quad \text{for } (t, x) \in \sigma.$$

The monotonicity requirement on f^i is redundant.

Consider the system of equations

$$\begin{cases} \mathcal{L}_c^i[u] - \frac{\partial u}{\partial t} := \sum_{j,k=1}^m a_{jk}^i(t, x) \frac{\partial^2 u}{\partial x_j \partial x_k} + \sum_{j=1}^m b_j^i(x) \frac{\partial u}{\partial x_j} + c^i(x)u - \frac{\partial u}{\partial t}, \\ \frac{\partial w}{\partial t} = g(t, x, u, w) \quad \text{for } (t, x) \in (0, T]. \end{cases}$$

Nevertheless, the monotonicity assumptions on the right-hand sides of this system are essential. The details follow:

Considering the special system

$$\begin{cases} \dfrac{\partial^2 u}{\partial x^2} - \dfrac{\partial u}{\partial t} = au + bw, \\ \dfrac{\partial w}{\partial t} = au + bw \quad \text{for } |x| < \dfrac{\pi}{2}, \end{cases}$$

it is shown by means of an example that the above comparison property need not hold in the absence of assumption $a \geq 0$ and $b \leq 0$.

If the diffusion process involves a larger number of the density functions $z^1, z^2, \ldots, z^r$ where $z^i = z^i(t, x), i = 1, 2, \ldots, r$, and this process admits the effect of convection, then we obtain the system of coupled *reaction-diffusion-convection equations*, which can be written in the form

$$\frac{\partial z^i}{\partial t} - \sum_{j,k=1}^{m} \frac{\partial}{\partial x_j}\left(d^i_{jk}(t,x)\frac{\partial z^i}{\partial x_k}\right) + \sum_{j=1}^{m} b^i_j(t,x)\frac{\partial z^i}{\partial x_j}$$

$$= f^i(t,x,z^1,\ldots,z^r), \quad i = 1, 2, \ldots, r, \tag{12.39}$$

where d^i_{jk} are coefficients of *multicomponent diffusion* and $\vec{b}^i := \vec{b}^i(t,x) = \left(b^i_1(t,x), \ldots, b^i_m(t,x)\right), i = 1, 2, \ldots, r$, is a drift vector.

The more *general reaction-diffusion-convection operators* of the form

$$\mathcal{A}^i[z^i](t,x) := \sum_{j,k=1}^{m} \frac{\partial}{\partial x_j}\left(d^i_{jk}(t,x)\frac{\partial z^i}{\partial x_k} + b^i_j(t,x)z^i\right)$$

$$- \sum_{j=1}^{m} d^i_j(t,x)\frac{\partial z^i}{\partial x_j} - d^i_0(t,x)z^i, \quad i \in S,$$

and suitable infinite uncountable systems of equations as a continuous model of coagulation-fragmentation processes with diffusion and the existence and uniqueness have been proved in the class of volume preserving solutions.

12.5 Monotone Iterative Methods for Finite Systems

Monotone iterative methods apply to studying the existence and uniqueness of the global classical solutions of initial-boundary value problems with both Dirichlet and nonlinear Neumann conditions. The results obtained have been applied to studying the existence of minimal and maximal solutions of the corresponding quasilinear reaction-diffusion equations arising frequently in neuroscience, with the differential operator in the divergence form (12.4)

$$\frac{\partial z}{\partial t} - \sum_{j,k=1}^{m} \frac{\partial}{\partial x_j}\left(a_{jk}(t,x,z)\frac{\partial z}{\partial x_k}\right) = f\left(t,x,z,\frac{\partial z}{\partial x}\right)$$

12.6 Extension of Monotone Iterative Methods to Infinite Systems

The monotone iterative methods can be extended to infinite systems of equations (with functional dependence of the right-hand sides of the system on the unknown function z and on spatial derivatives $\mathcal{D}_x z^i$) of the form

$$\mathcal{D}_t z^i(t, x) - \sum_{j,k=1}^{m} a_{jk}^i \Big(t, x, z^i(t, x), \mathcal{D}_x z^i(t, x)\Big) \mathcal{D}_{x_j x_k}^2 z^i(t, x)$$

$$= f^i\Big(t, x, z(t, x), \mathcal{D}_x z^i(t, x), z(\cdot, \cdot), \mathcal{D}_x z^i(\cdot, \cdot)\Big), \quad i \in S,$$

(12.40)

where

$$z(\cdot, \cdot) := z, \quad \mathcal{D}_x z^i(\cdot, \cdot) := \mathcal{D}_x z^i$$

with classical or nonlinear nonlocal initial-boundary conditions.

Techniques and methods used in Part I may be used to construct iterative methods for infinite systems of these reaction-diffusion equations with functionals.

12.7 Conclusions

Reaction-diffusion equations describe a variety of phenomena in neuroscience. Physical models of the reaction-diffusion equations include: (i) the propagation of an action potential along an electrical cable (Hodgkin-Huxley equations); (ii) concentration changes of ionic species under the influence of local reactions, and diffusion of ions down their concentration gradients without expense of energy in an aqueous medium (Kolmogorov-Petrovsky-Piskunov equations); and (iii) electrodiffusion – a nonlinear transport process whose essence is diffusion of ions combined with their migration in an electric field (Nernst-Planck equations). However, it is only in the presence of a large number of molecules or ions that electrodiffusion is the principal means of migration. It is the understanding of particular electrodiffusional transport phenomena that is necessary for a comprehensive picture. So far, this picture is incomplete. Electrodiffusion has not yet been reconciled with action potential propagation and alternative ways (not reaction-diffusion systems, but those based on Maxwell's equations) are appearing in theoretical models of information handling in the brain.

A.1 List of Symbols Used for Approximation Sequences

I. $\{u_n\}, \{v_n\}$ – Method of Direct Iterations

$$\mathcal{F}^i[u_n^i](t, x) = f^i(t, x, u_{n-1}(t, x), u_{n-1}), \tag{A.1.1}$$

$$\mathcal{F}^i[v_n^i](t, x) = f^i(t, x, v_{n-1}(t, x), v_{n-1}), \quad i \in S. \tag{A.1.2}$$

II. $\{\hat{u}_n\}, \{\hat{v}_n\}$ – Chaplygin Method

$$\mathcal{F}^i[\hat{u}_n^i](t, x) = f^i(t, x, \hat{u}_{n-1}(t, x), \hat{u}_{n-1}) + f_{y^i}^i(t, x, \hat{u}_{n-1}(t, x), \hat{u}_{n-1}) \tag{A.1.3}$$
$$\cdot [\hat{u}_n^i(t, x) - \hat{u}_{n-1}^i(t, x)],$$

$$\mathcal{F}^i[\hat{v}_n^i](t, x) = f^i(t, x, \hat{v}_{n-1}(t, x), \hat{v}_{n-1}) + f_{y^i}^i(t, x, \hat{v}_{n-1}(t, x), \hat{v}_{n-1}) \tag{A.1.4}$$
$$\cdot [\hat{v}_n^i(t, x) - \hat{v}_{n-1}^i(t, x)], \quad i \in S.$$

III. $\{\overset{*}{u}_n\}, \{\overset{*}{v}_n\}$ – First Variant of Chaplygin Method

$$\mathcal{F}_k^i[\overset{*}{u}_n^i](t, x) = f^i(t, x, \overset{*}{u}_{n-1}(t, x), \overset{*}{u}_{n-1}) + k^i(t, x)\overset{*}{u}_{n-1}^i(t, x), \tag{A.1.5}$$

$$\mathcal{F}_k^i[\overset{*}{v}_n^i](t, x) = f^i(t, x, \overset{*}{v}_{n-1}(t, x), \overset{*}{v}_{n-1}) + k^i(t, x)\overset{*}{v}_{n-1}^i(t, x), \quad i \in S. \tag{A.1.6}$$

IV. $\{\overset{\circ}{u}_n\}, \{\overset{\circ}{v}_n\}$ – Second Variant of Chaplygin Method

$$\mathcal{F}_\kappa^i[\overset{\circ}{u}_n^i](t, x) = f^i(t, x, \overset{\circ}{u}_{n-1}(t, x), \overset{\circ}{u}_{n-1}) + \kappa \overset{\circ}{u}_{n-1}^i(t, x), \tag{A.1.7}$$

$$\mathcal{F}_\kappa^i[\overset{\circ}{v}_n^i](t, x) = f^i(t, x, \overset{\circ}{v}_{n-1}(t, x), \overset{\circ}{v}_{n-1}) + \kappa \overset{\circ}{v}_{n-1}^i(t, x), \quad i \in S. \tag{A.1.8}$$

V. $\{\tilde{u}_n\}, \{\tilde{v}_n\}$ – Wazewski Method

$$\mathcal{F}^i[\tilde{u}_n^i](t, x) = f^i(t, x, \tilde{u}_n(t, x), \tilde{u}_{n-1}), \tag{A.1.9}$$

$$\mathcal{F}[\tilde{v}_n^i](t, x) = f^i(t, x, \tilde{v}_n(t, x), \tilde{v}_{n-1}), \quad i \in S. \tag{A.1.10}$$

VI. $\{\overline{u}_n\}, \{\overline{v}_n\}$ – Mlak-Olech Method

$$\mathcal{F}^i[\overline{u}_n^i](t, x) = f^i(t, x, [\overline{u}_n, \overline{u}_{n-1}]^i) \tag{A.1.11}$$
$$:= f^i(t, x, \overline{u}_{n-1}^1, \dots, \overline{u}_{n-1}^{i-1}, \overline{u}_n^i, \overline{u}_{n-1}^{i+1}, \dots),$$

$$\mathcal{F}^i[\overline{v}_n^i](t, x) = f^i(t, x, [\overline{v}_n, \overline{v}_{n-1}]^i) \tag{A.1.12}$$
$$:= f^i(t, x, \overline{v}_{n-1}^1, \dots, \overline{v}_{n-1}^{i-1}, \overline{v}_n^i, \overline{v}_{n-1}^{i+1}, \dots), \quad i \in \mathbb{N}.$$

Mathematical Neuroscience. http://dx.doi.org/10.1016/B978-0-12-411468-5.00020-X

A.2 Existence and Uniqueness Theorems

Let D be a bounded cylindrical domain $D := (0, T] \times G$, where $0 < T < \infty$ and $G \subset \mathbb{R}^m$ is an open bounded domain with boundary $\partial G \in C^{2+\alpha}$ $(0 < \alpha < 1)$.

Let us consider the *Fourier first initial-boundary value problem* for an infinite system of linear parabolic equations

$$\begin{cases} \mathcal{F}_c^i[u^i](t, x) = g^i(t, x), & i \in S, \quad \text{for } (t, x) \in D, \\ u(0, x) = \phi_0(x) \quad \text{for } x \in G, \\ u(t, x) = h(t, x) \quad \text{for } (x, t) \in \sigma \end{cases} \tag{A.2.1}$$

with the compatibility condition

$$h(0, x) = \phi_0(x) \quad \text{for } x \in \partial G,$$

where

$$\mathcal{F}_c^i := \mathcal{D}_t - \mathcal{L}_c^i, \quad \mathcal{L}_c^i := \sum_{j,k=1}^m a_{jk}^i(t, x)\mathcal{D}_{x_j x_k}^2 - \sum_{j=1}^m b_j^i(t, x)\mathcal{D}_{x_j} - c^i(t, x)\mathcal{I},$$

$(t, x) \in D$, $\mathcal{I}$ is the identity operator, the operators $\mathcal{F}_c^i$, $i \in S$, are uniformly parabolic in $\overline{D}$, and s is a set of indices (finite or infinite).

From the theorems on the existence and uniqueness of solutions of the Fourier first initial-boundary value problems for linear parabolic equations in Hölder spaces, we directly infer the following theorem:

Theorem A.1 *Let us consider linear parabolic initial-boundary value problem* (A.2.1) *and assume that*

i. *all the coefficients a_{jk}^i, b_j^i, and c^i $(j, k = 1, \ldots, m, i \in S)$ of the operators $\mathcal{L}_c^i$, $i \in S$, fulfill assumption (H_c);*
ii. *the functions g^i, $i \in S$, are uniformly Hölder continuous with respect to t and x in $\overline{D}$, i.e.,*
  $g = \{g^i\}_{i \in s} \in C_S^{0+\alpha}(\overline{D});$
iii. $\phi_0 \in C_S^{2+\alpha}(G), h \in C_S^{2+\alpha}(\sigma),$ *and* $\partial G \in C_S^{2+\alpha}.$

Then problem (A.2.1) *has the unique solution u and $u \in C_S^{2+\alpha}(\overline{D})$.*

Moreover, there exists a constant $C > 0$ depending only on the constants μ, K_1, α, and on the geometry of the domain D, such that the following a priori Schauder estimate of the $(2 + \alpha)$-type holds

$$\|u\|_{2+\alpha} \leq C \left(\|g\|_{0+\alpha} + \|\phi\|_{2+\alpha}^G + \|h\|_{2+\alpha}^\sigma \right). \tag{A.2.2}$$

Proof. Observe that system (A.2.1) has the following property: in the ith equation, only one unknown function with index i appears. Therefore, system (A.2.1) is a collection of individual independent equations. Applying the above-mentioned theorems, we immediately obtain the estimates

$$|u^i|_{2+\alpha} \le C_i \left(|g^i|_{0+\alpha} + |\phi_0^i|_{2+\alpha}^G + |h^i|_{2+\alpha}^\sigma \right), \quad i \in S$$

in which constants $C_i > 0$ are independent of g^i, ϕ_0^i, and h^i.

From these theorems it follows that constants C_i depend on the constants μ, K_1, α only and on the geometry of the domain D; C_i are also uniformly bounded for all $i \in S$. Therefore, there exists a constant $C > 0$ (C is independent of both the index i and the functions g^i, ϕ_0^i, and h^i) such that $C_i \le C$, for all $i \in S$. Hence, by the definitions of the norms in the spaces $C_S^{0+\alpha}(\overline{D})$ and $C_S^{2+\alpha}(\overline{D})$, we obtain estimate (A.2.2). $\qquad\square$

Let us consider the homogeneous first initial-boundary value problem

$$\begin{cases} \mathcal{F}_c^i[u^i](t, x) = g^i(t, x), \quad i \in S, \quad \text{for } (t, x) \in D, \\ u(0, x) = 0 \quad \text{for } x \in G, \\ u(t, x) = 0 \quad \text{for } (t, x) \in \sigma. \end{cases} \quad (A.2.3)$$

Using the same arguments as previously, we directly get the following theorem.

Theorem A.2 *Assume that $g = \{g^i\}_{i \in S} \in C_S(\overline{D})$, $\partial G \in C^{2+\alpha} \cap C^{2-0}$, condition (H_a) hold and $a_{jk}^i \in C^{1-0}$ on σ. Let $g(0, x)$ vanish on ∂G and u be a solution of problem (A.2.3).*

Then, for any $\beta, 0 < \beta < 1$, there exists a constant $K > 0$ depending only on β, u_0, K_1, K_2, and the geometry of the domain D such that

$$\|u\|_{1+\beta} \le K\|g\|_0. \quad (A.2.4)$$

Moreover, there exists a constant $\overline{K} > 0$ depending on the same parameters as K and such that

$$\|u\|_{1+\beta}^{D\tau} \le \overline{K} \tau^{\frac{1-\beta}{2}} \|g\|_0^{D\tau} \quad (A.2.5)$$

for sufficiently small τ, i.e., $2\overline{K}\tau^{\frac{1-\beta}{2}} \le 1 \ (0 < \tau \le T)$.

A.3 Integral Representations of Solutions

Let us consider the Fourier first initial-boundary value problem in a bounded cylindrical domain D for the infinite system of equations

$$\begin{cases} \mathcal{F}_c[u](t, x) = g(t, x) \quad \text{for } (t, x) \in D, \\ u(0, x) = \phi_0(x) \quad \text{for } x \in G, \\ u(t, x) = h(t, x) \quad \text{for } (t, x) \in \sigma \end{cases} \quad (A.3.1)$$

with the compatibility condition

$$h(0, x) = \phi_0(x) \quad \text{for } x \in \partial G,$$

where

$$\mathcal{F}_c := \mathcal{D}_t - \mathcal{L}_c, \quad \mathcal{L}_c := \sum_{j,k=1}^{m} a_{jk}(t, x)\mathcal{D}^2_{x_j x_k} - \sum_{j=1}^{m} b_j(t, x)\mathcal{D}_{x_j} - c(t, x)\mathcal{I},$$

and the operator $\mathcal{F}_c$ is uniformly parabolic in $\overline{D}$.

From the theorems on the unique solvability of the Fourier first initial-boundary value problems, we get the following theorem:

Theorem A.3   *If all coefficients a_{jk}, b_j, and c of the operator $\mathcal{L}_c$ belong to the class $C^{0+\alpha}(\overline{D})$ and in addition coefficients a_{jk} have derivatives $\mathcal{D}_x a_{jk}$ satisfying the Hölder condition with respect to x with the exponent α $(0 < \alpha < 1)$, the function ψ is continuous and $\partial G \in C^{2+\alpha}$, then problem (A.3.1) with $g(t, x) = 0$ and $\phi_0(x) = 0$, i.e., the problem*

$$\begin{cases} \mathcal{F}_c[u](t, x) = 0 & \text{for } (t, x) \in D, \\ u(0, x) = 0 & \text{for } x \in G, \\ u(t, x) = h(t, x) & \text{for } (t, x) \in \sigma, \end{cases} \tag{A.3.2}$$

where $h(0, x) = 0$ on ∂G, has the unique regular solution u in $\overline{D}$ given by the integral formula

$$u = u_1(t, x) = \int_0^t \int_{\partial G} \frac{d\Gamma}{dv_\xi}(t, x; \tau, \xi)\Psi(\tau, \xi)d\tau d\sigma. \tag{A.3.3}$$

This integral is known as a double-layer potential with a density Ψ which is governed by the following Volterra integral equation:

$$\Psi(t, x) = 2\int_0^t \int_{\partial G} \frac{d\Gamma}{dv_\xi}(t, x; \tau, \xi)\Psi(\tau, \xi)d\tau d\sigma - 2h(t, x) \quad \text{in } D. \tag{A.3.4}$$

$\Gamma(t, x; \tau, \xi)$ *is the fundamental solution of the reaction-diffusion equation*

$$\mathcal{D}_t u(t, x) - \mathcal{L}[u](t, x) = 0 \quad \text{in } D,$$

where $\frac{d}{dv_\xi}$ is the outward normal derivative on ∂G with respect to the variable ξ and $d\sigma$ is the surface element of σ.

Theorem A.4   *If all coefficients a_{jk}, b_j, and c of the operator $\mathcal{L}_c$ belong to the class $C^{0+\alpha}(\overline{D})$, the functions g and ϕ_0 are bounded, the function g is locally Hölder continuous with respect to x uniformly with respect to l in $\overline{D}$, the function ϕ_0 is continuous and satisfies* the compatibility condition $\phi_0(x) = 0$ on ∂G and $g(0, x) = 0$ on ∂G, then problem (A.3.1) for $h(t, x) = 0$, i.e., the problem

$$\begin{cases} \mathcal{F}_c[u](t, x) = g(t, x) & for\ (t, x) \in D, \\ u(0, x) = \phi_0(x) & for\ x \in G, \\ u(t, x) = 0 & for\ (t, x) \in \sigma, \end{cases} \tag{A.3.5}$$

has the unique regular solution u in the domain $\overline{D}$ given by the following integral formula:

$$\begin{aligned} u &= u_2(t, x) + u_3(t, x) \\ &= \int_0^t \int_G \mathcal{G}(t, x; \tau, \xi)g(\tau, \xi)d\tau d\xi + \int_G \mathcal{G}(t, x; 0, \xi)\phi_0(\xi)d\xi, \end{aligned} \tag{A.3.6}$$

where $\mathcal{G}(t, x; \tau, \xi)$ is Green's function for initial-boundary value problem (A.3.5).

From the theorems on the existence of the fundamental solution, the theorem follows:

Theorem A.5 *If all coefficients a_{jk}, b_j, and c of the operator $\mathcal{L}_c$ satisfy the condition (H_c) in $\overline{D}$, then there exists the fundamental solution $\Gamma(t, x; \tau, \xi)$ of the linear equation*

$$\mathcal{D}_t u(t, x) - \mathcal{L}_c[u](t, x) = 0 \quad in\ D. \tag{A.3.7}$$

Corollary A.1 *If all assumptions of Theorems A.3 and A.4 hold, then the problem (A.3.1) has the unique regular solution u in $\overline{D}$, and u is given by the following integral formula:*

$$\begin{aligned} u &= u_1(t, x) + u_2(t, x) + u_3(t, x) \\ &= \int_0^t \int_{\partial G} \frac{d\Gamma}{d\nu_\xi}(t, x; \tau, \xi)\Psi(\tau, \xi)d\tau d\sigma + \int_0^t \int_G \mathcal{G}(t, x; \tau, \xi)g(\tau, \xi)d\tau d\xi \\ &\quad + \int_G \mathcal{G}(t, x; 0, \xi)\phi_0(\xi)d\xi, \end{aligned} \tag{A.3.8}$$

where the density Ψ is defined by (A.3.4).

Corollary A.2 *Let all assumptions of Theorem A.1 hold. Then the integral representations of the solution of problem (A.2.1) are given by the following integral formulas:*

$$\begin{aligned} u^i &= u^i(t, x) \\ &= \int_0^t \int_{\partial G} \frac{d\Gamma^i}{d\nu_\xi}(t, x; \tau, \xi)\Psi^i(\tau, \xi)d\tau d\sigma + \int_0^t \int_G \mathcal{G}^i(t, x; \tau, \xi)g^i(\tau, \xi)d\tau d\xi \\ &\quad + \int_G \mathcal{G}^i(t, x; 0, \xi)\phi_0^i(\xi)d\xi, \quad i \in S, \end{aligned} \tag{A.3.9}$$

where, for each $i \in S$, the functions Γ^i and $\mathcal{G}^i$ are just the same as for suitable scalar parabolic-reaction-diffusion problem (A.3.1) and the densities Ψ^i, for $i \in S$, are governed by *the following Volterra integral equations:*

$$\Psi^i(t, x) = 2 \int_0^t \int_{\partial G} \frac{d\Gamma^i}{d\upsilon_\xi}(t, x; \tau, \xi)\Psi^i(\tau, \xi)d\tau d\sigma - 2h^i(t, x) \quad in \ D. \tag{A.3.10}$$

A.4 Weak C-Solution

A function $z \in C_S(\overline{D})$ is called a weak C-solution of the initial-boundary value problem

$$\begin{cases} \mathcal{F}^i[z^i](t, x) = f^i(t, x, z(t, x), z) & \text{for } (t, x) \in D, \\ z^i(0, x) = \phi_0^i(x) & \text{for } x \in G, \\ z^i(t, x) = h^i(t, x) & \text{for } (t, x) \in \sigma \ \text{ and } \ i \in S \end{cases} \tag{A.4.1}$$

with the compatibility condition

$$h(0, x) = \phi_0(x) \quad \text{for } x \in \partial G,$$

if z satisfies the following infinite system of the integral equations:

$$z^i(l, x) = \int_0^t \int_{\partial G} \frac{d\Gamma^i}{d\upsilon_\xi}(l, x; \tau, \xi)\Psi^i(\tau, \xi)d\tau d\sigma$$

$$+ \int_0^t \int_G \mathcal{G}^i(l, x; \tau, \xi)f^i(\tau, \xi, z(\tau, \xi), z)d\tau d\xi \tag{A.4.2}$$

$$+ \int_G \mathcal{G}^i(t, x; 0, \xi)\phi_0^i(\xi)d\xi \quad \text{for } (t, x) \in \overline{D}, \quad i \in S$$

for $(t, x) \in \overline{D}$, where the densities Ψ^i, $i \in S$, are governed by integro-differential equations (A.3.10).

In this sense differential problem (A.4.1) is equivalent to the infinite system of integral equations (A.4.2).

A.5 Integral Transformation

We define the transformation **T** setting

$$\mathbf{T}: C_S(\overline{D}) \to C_S(\overline{D}), \quad \beta \mapsto \mathbf{T}[\beta] = \gamma,$$

where γ is the (supposedly unique) weak C-solution of the problem

$$\begin{cases} \mathcal{F}^i[\gamma^i](t, x) = f^i(t, x, \beta(t, x), \beta) & \text{for } (t, x) \in D, \\ \gamma^i(0, x) = \phi_0^i(x) & \text{for } x \in G, \\ \gamma^i(t, x) = h^i(t, x) & \text{for } (t, x) \in \sigma \ \text{ and } \ i \in S, \end{cases} \tag{A.5.1}$$

which $\gamma^i, i \in S$, are given by the integral formulae

$$
\gamma^i(t, x) = \int_0^t \int_{\partial G} \frac{d\Gamma^i}{dv_\xi}(t, x; \tau, \xi)\Psi^i(\tau, \xi)d\tau d\sigma
$$

$$
+ \int_0^t \int_G \mathcal{G}^i(t, x; \tau, \xi)f^i(\tau, \xi, \beta(\tau, \xi), \beta)d\tau d\xi \tag{A.5.2}
$$

$$
+ \int_G \mathcal{G}^i(t, x; 0, \xi)\phi_0^i(\xi)d\xi
$$

for $(t, x) \in \overline{D}, i \in S$, where the densities $\Psi^i, i \in S$, are governed by infinite system of Voltorra integral equations (A.3.10).

Therefore, the problem of solving initial-boundary value problem (A.4.1) is in some sense equivalent to the fixed point problem for the transformation of **T**, and (A.5.2) are called fixed point equations.

Further Reading

Introductory knowledge of linear functional analysis is essential for an understanding of concepts used throughout this book (see e.g. Kreyszig, 1978). Nonlinear analysis of partial differential equations based on functional analytic methods can be found in monographs by Friedman (1964, 2008). Recent introductory volumes on mathematical neuroscience by Scott (2002) and Ermentrout and Terman (2010) complement exposition of mathematical neuroscience based on classical mathematics.

In the theory of differential inequalities, the monographs by Lakshmikantham and Leela (1969) and Szarski (1965) play a crucial role. The papers of Walter (1997, 2002) give a review of the problems in the theory of parabolic equations and inequalities. In the case of finite systems of differential inequalites, the fundamental results were obtained by Szarski (1973, 1974, 1975, 1976) and, under somewhat different assumptions, by Redheffer and Walter (1976, 1977, 1979, 1980) and Redlinger (1984, 1988). In the case of infinite systems of differential inequalities, the fundamental results were obtained by Szarski (1979, 1980a,b), Jaruszewska-Walczak (2001, 2005), Kamont (2002), Kraśnicka (1982, 1987), and Brzychczy (2004).

Numerous examples of finite systems of equations are given by Diekmann and Temme (1982), Fife (1979), Leung (1989), Logan (1994), Pao (1974, 1976, 1992, 1995, 1998), Rothe (1984), Smoller (1983), and Wu (1996).

Monotone iterative methods have been studied by numerous authors: Amann (1971, 1975, 1976), Bange (1974), Brzychczy (1963a,b, 1965, 1983, 1986, 1987, 1989a,b, 1993, 1995a,b, 1996, 1999), Bychowska (2004), Diekmann and Temme (1982), Kusano (1965), Leszczyński (1998a,b, 1999), Keller (1969), čojczyk-Krolikiewicz (1993), Mikhlin and Smolickiǐ (1965), Mlak (1960), Mlak and Olech (1963), Mysovskikh (1954), Nowotarska (1975), Sattinger (1972, 1973), Smoller (1983), Vatsala and Yang (2006), Zeragia (1957, 1973, 1956); and for problems with nonlocal boundary and nonlocal initial conditions by, e.g., Carl and Heikkilä (2000), Pao (1995, 1998). Abstract monotone iterative methods in ordered Banach spaces have been studied by Liz (1977), Liz and Nieto (1998). We should also note that the monographs by Ladde et al. (1985) and Pao (1992) play a crucial role in this field.

Mathematical Neuroscience. http://dx.doi.org/10.1016/B978-0-12-411468-5.00021-1

Infinite systems equations have been studied by numerous authors, including: Banaś and Lecko (2001a,b, 2002) and Banaś and Sadarangani (2003), Brzychczy (1999, 2000a,b, 2001, 2002a,b,c, 2004, 2005), Chandra et al. (1978), Deimling (1977), Du and Lakshmikantham (1982), Jaruszewska-Walczak (2001, 2005), Kamont (2003), Kamont and Kozieş (2003), Mlak and Olech (1963), Moszyński and Pokrzywa (1974), Persidskiĭ (1958, 1959, 1960), Pogorzelski (1955, 1957), Pudełko (2002, 2004, 2005a,b, 2007), Rzepecki (1975, 1977), Zabawa (2005, 2006), Zhautykov (1965), and Valeev and Zhautykov (1974).

The Chaplygin method has been introduced by Chaplygin (1950) for ordinary differential equations and developed by Lusin (1953). Zeragia (1956, 1957, 1973), Mlak (1960) and Brzychczy (1963, 1965, 1983); Brzychczy et al. (1974) have subsequently applied this method to parabolic partial differential equations and Mysovskikh (1954) to elliptic equations, and Zeragia (1965) to hyperbolic equations. In this method, the rate of convergence is quadratic (i.e., Newton speed). The Chaplygin method has several variants in the literature and appears there under different names: Newton-Kantorovitch's method (see Mikhlin and Smolickiĭ, 1965), Zeragia's method (see Rabczuk, 1976), quasilinearization method (see Carl and Lakshmikantham, 2002), or Bellman-Kalaba-Lakshmikantham quasilinearization method (see Ahmad et al., 2001). Using Chaplygin's idea, Bellman and Kalaba (1965) developed this method as a method of quasilinearization and, following the publication of numerous interesting papers by Lakshmikantham (1994, 2005) as a generalized quasilinearization method. The Chaplygin method is in some sense an extension of the Newton method of solving nonlinear algebraic equations onto differential equations (cf. Lusin (1953), Lakshmikantham and Vatsala (2005). In particular the Chaplygin method has been extended and applied in numerous papers by Kamont (1980, 1999), Leszcznýski (1998a,b, 1999, 2000), Czşapiński (1999), Jaruszewska-Walczak, (1999, 2001, 2004), Bychowska (2002, 2004, and Bartşomiejczyk and Leszczyński, 2004).

Another variant of the Chaplygin method has frequently been applied by several authors Amann (1976), Diekmann and Temme (1982), Pao (1976, 1992), Sattinger (1972, 1973), and Smoller (1983).

References

This list includes references that will be useful for further study.

Adams, R. A., *Sobolev Spaces*, Academic Press, New York, 1975.

Agmon, S., Douglis, A. and Nirenberg, L., *Estimates near the boundary for solutions of elliptic partial differential equations satisfying general boundary conditions*, Part I, Commun. Pure Appl. Math., 12 (1959), pp. 623–727.

Ahmad, B., Nieto, J. J. and Shahzad, N., *The Bellman-Kalaba-Lakshmikantham quazilinearization method for Neumann problems*, J. Math. Anal. Appl., 257 (2001), pp. 356–363.

Amann, H., *On the existence of positive solutions of nonlinear elliptic boundary value problems*, Indiana Univ. Math. J., 21 (1971), pp. 125–146.

Amann, H., *Nonlinear operators in ordered Banach spaces and some applications to nonlinear boundary value problems*. In: *Nonlinear Operators and the Calculus of Variations*, Lecture Notes in Mathematics, Vol. 543, Springer-Verlag, Berlin, 1975, pp. 1–55.

Amann, H., *Supersolutions, monotone iterations and stability*, J. Differential Equations, 21 (1976), pp. 363–377.

Amann, H., *Coagulation-fragmentation processes*, Arch. Ration. Mech. Anal., 151 (2000), pp. 339–366.

Amari, S.-I., *Dreaming of mathematical neuroscience for half a century*, Neural Netw., 37 (2012), pp. 48–51.

Andres, J. and Górniewicz, L., *Topological Fixed Point Principles for Boundary Value Problems*, Kluwer Academic Publishers, Dordrecht, 2003.

Appell, J. and Zabrejko, P. P., *Nonlinear Superposition Operators*, Cambridge Tracts in Mathematics, Vol. 95, Cambridge University Press, Cambridge, 1990.

Aronson, D. G. and Weinberger, H. F., *Nonlinear diffusion in population genetics, combustion, and nerve pulse propagation*. In: J. A. Goldstein (Ed.), Lecture Notes in Mathematics, Vol. 446, Springer-Verlag, New York, 1975.

Ball, J. M. and Carr, J., *The discrete coagulation-fragmentation equations: existence, uniqueness and density conservation*, J. Stat. Phys., 61 (1990), pp. 203–234.

Banaś, J. and Lecko, M., *An existence theorem for a class of infinite systems of integral equations*, Math. Comput. Modelling, 34 (2001a), pp. 533–539.

Banaś, J. and Lecko, M., *Solvability of infinite systems of differential equations in Banach sequence space*, J. Comput. Appl. Math., 137 (2001b), pp. 363–375.

Banaś, J. and Lecko, M., *On solutions of an infinite system of differential equations*, Dyn. Syst. Appl., 11 (2002), pp. 221–230.

Banaś, J. and Sadarangani, K., *Solutions of some functional-integral equations in Banach algebra*, Math. Comput. Modelling, 38 (2003), pp. 245–250.

Bange, D., *A constructive existence theorem for a nonlinear parabolic equation*, SIAM J. Math. Anal., 5 (1974), pp. 103–110.

Bartłomiejczyk, A. and Leszczyński, H., *Comparison principles for parabolic differential-functional initial-value problems*, Nonlinear Anal., 57 (2004), pp. 63–84.

Beckenbach, E. F. and Bellman, R., *Inequalities*, Springer-Verlag, Berlin, 1961.

Bell, J., *Some threshold results for models of myelinated nerves*, Math. Biosci., 54 (1981), pp. 181–190.

Bell, J. and Cosner, C., *Threshold behavior and propagation for nonlinear differential-difference systems motivated by modeling myelinated axons*, Quart. Appl. Math., 42 (1984), pp. 1–14.

Mathematical Neuroscience. http://dx.doi.org/10.1016/B978-0-12-411468-5.00022-3

Bellman, R. and Kalaba, R., *Quasilinearization and Nonlinear Boundary Value Problems*, Elsevier, New York, 1965.

Bellout, H., *Blow-up of solutions of parabolic equations with nonlinear memory*, J. Differential Equations, 70 (1978), pp. 42–68.

Bénilan, Ph. and Wrzosek, D., *On an infinite system of reaction-diffusion equations*, Adv. Math. Sci. Appl., 7 (1997), pp. 349–364.

Besala, P., *On solutions of Fourier's first problem for a system of non-linear parabolic equations in an unbounded domain*, Ann. Polon. Math., 13 (1963), pp. 247–265.

Besala, P., *An extension of the strong maximum principle for parabolic equations*, Bull. Acad. Polon. Sci. Math. Astr. Phys., 19 (1971), pp. 1003–1006.

Bicadze, A. V. and Samarski, A. A., *On some simple generalizations of linear elliptic boundary problems*, Soviet Math. Dokl., 10 (1969), pp. 398–400.

Brzychczy, S., *Extension of Chaplygin's method to the system of nonlinear parabolic equations in an unbounded domain*, Ph.D. Thesis, Jagiellonian University, Cracow, 1963 [Polish].

Brzychczy, S., *Some theorems on second order partial differential inequalities of parabolic type*, Ann. Polon. Math., 15 (1964), pp. 143–151.

Brzychczy, S., *Extension of Chaplygin's method to the system of nonlinear parabolic equations in an unbounded domain*, Bull. Acad. Polon. Sci. Math. Astr. Phys., 13 (1965), pp. 27–30.

Brzychczy, S., *On a certain property of a transformation in Hilbert space in connection with the theory of differential equation*, Colloq. Math., 18 (1967), pp. 143–146.

Brzychczy, S., *A comparison of solutions of Prandtl's and Navier-Stokes's systems of equations at low viscosity*, Bull. Acad. Polon. Sci. Math. Astr. Phys., 16 (1968), pp. 175–180 [Russian].

Brzychczy, S., *Approximate iterative method and the existence of solutions of non-linear parabolic differential-functional equations*, Ann. Polon. Math., 42 (1983), pp. 37–43.

Brzychczy, S., *Chaplygin's method for a system of nonlinear parabolic differential-functional equations*, Differ. Uravn., 22 (1986), pp. 705–708 [Russian].

Brzychczy, S., *Existence of solution for nonlinear systems of differential-functional equations of parabolic type in an arbitrary domain*, Ann. Polon. Math., 47 (1987), pp. 309–317.

Brzychczy, S., *An estimate for the rate of convergence of successive Chaplygin approximations for a parabolic system of functional-differential equations*, Differ. Uravn., 25 (1989a), pp. 1050–1052 [Russian].

Brzychczy, S., *On an estimate for the modulus of the solution of a nonlinear system of functional-differential equations of parabolic type*, Differ. Uravn., 25 (1989b), pp. 1444–1446 [Russian].

Brzychczy, S., *Existence of solution of the nonlinear Dirichlet problem for differential-functional equations of elliptic type*, Ann. Polon. Math., 58 (1993), pp. 139–146.

Brzychczy, S., *On a certain approximate method for nonlinear system of differential-functional equations of parabolic type*, Opuscula Math., 15 (1995a), pp. 45–50.

Brzychczy, S., *Monotone Iterative Methods for Nonlinear Parabolic and Elliptic Differential-Functional Equations*, Dissertations Monographies, Vol. 20, University of Mining and Metallurgy Publishers, Cracow, 1995b.

Brzychczy, S., *On the stability of solutions of nonlinear parabolic differential-functional equations*, Ann. Polon. Math., 63 (1996), pp. 155–165.

Brzychczy, S., *Existence of solutions and monotone iterative method for infinite systems of parabolic differential-functional equations*, Ann. Polon. Math., 72 (1999), pp. 15–24.

Brzychczy, S., *Chaplygin's method for infinite systems of parabolic differential-functional equations*, Univ. Iagel. Acta Math., 38 (2000a), pp. 153–162.

Brzychczy, S., *Some variant of iteration method for infinite systems of parabolic differential-functional equations*, Opuscula Math., 20 (2000b), pp. 41–50.

Brzychczy, S., *Existence and uniqueness of solutions of nonlinear infinite systems of parabolic differential-functional equations*, Ann. Polon. Math., 77 (2001), pp. 1–9.

Brzychczy, S., *On the existence of solutions of nonlinear infinite systems of parabolic differential-functional equations*, Univ. Iagel. Acta Math., 40 (2002a), pp. 31–38.

Brzychczy, S., *Existence of solutions of nonlinear infinite systems of parabolic differential-functional equations*, Math. Comput. Modelling, 36 (2002b), pp. 435–443.

Brzychczy, S., *Existence and uniqueness of solutions of infinite systems of semilinear parabolic differential-functional equations in arbitrary domains in ordered Banach spaces*, Math. Comput. Modelling, 36 (2002c), pp. 1183–1192.

Brzychczy, S., *Infinite systems of strong parabolic differential-functional inequalities*, Univ. Iagel. Acta Math., 42 (2004), pp. 139–148.

Brzychczy, S., *Monotone iterative methods for infinite systems of reaction-diffusion-convection equations with functional dependence*, Opuscula Math., 25 (2005), pp. 29–99.

Brzychczy, S., *Infinite Systems of Parabolic Differential-Functional Equations*, Monograph, AGH University of Science and Technology Press, Cracow, 2006.

Brzychczy, S. and Górniewicz, L., *Continuous and discrete models of neural systems in infinite-dimensional abstract spaces*, Neurocomputing, 74 (2011), pp. 2711–2715.

Brzychczy, S. and Janus, J., *On some monotone iterative method for nonlinear hyperbolic differential-functional equations*. Proceedings of the Sixth International Colloquium on Differential Equations, Plovdiv, Bulgaria, 1995, pp. 69–76.

Brzychczy, S. and Janus, J., *Monotone iterative methods for nonlinear integro-differential hyperbolic equations*, Univ. Iagel. Acta Math., 37 (1999), pp. 245–261.

Brzychczy, S. and Janus, J., *Monotone iterative methods for nonlinear hyperbolic integro-differential-functional equations*, Univ. Iagel. Acta Math., 38 (2000), pp. 141–152.

Brzychczy, S. and Poznanski, R. R., *Neuronal models in infinite-dimensional spaces and their finite-dimensional projections*, Part I, J. Integr. Neurosci., 9 (2010), pp. 11–30.

Brzychczy, S., Kapturkiewicz, W. and Węglowski, Z., *On application of certain difference schemes for determination of temperature distribution in the complex system: casting-metal mould*, Zesz. Nauk. AGH, Metal. Oldew, 60 (1974), pp. 65–72 [Polish].

Brzychczy, S., Leszczyński, H. and Poznanski, R. R., *Neuronal models in infinite-dimensional spaces and their finite-dimensional projections*. Part II, J. Integr. Neurosci., 11 (2012), pp. 265–276.

Busemeyer, J. R. and Bruza, P. D., *Quantum Models of Cognition and Decision*, Cambridge University Press, Cambridge, 2012.

Bychowska, A., *Quasilinearization methods for nonlinear parabolic equations with functional dependence*, Georgian Math. J., 9 (2002), pp. 431–448.

Bychowska, A., *Existence of unbounded solution to parabolic equations with functional dependence*, Math. Nachr., 263 (2004), pp. 53–66.

Cardanobile, S. and Mugnolo, D., *Analysis of a FitzHugh-Nagumo-Rall model of a neuronal network*. Math. Methods Appl. Sci., 30 (2007), pp. 2281–2308.

Carl, S. and Heikkilä, S., *Discontinuous reaction-diffusion equations under discontinuous and nonlocal flux conditions*, Math. Comput. Modelling, 32 (2000), pp. 1333–1344.

Carl, S. and Lakshmikantham, V., *Generalized quasilinearization method for reaction-diffusion equations under nonlinear and nonlocal flux conditions*, J. Math. Anal. Appl., 271 (2002), pp. 182–205.

Carpio, A. and Bonilla, L. L., *Pulse propagation in discrete systems of coupled excitable cells*, SIAM J. Appl. Math., 63 (2003), pp. 619–635.

Chabrowski, J., *On non-local problems for elliptic linear equations*, Funkcial. Ekvac., 32 (1989), pp. 215–226.

Chandra, J., Lakshmikantham, V. and Leela, S., *A monotone method for infinite systems of nonlinear boundary value problems*, Arch. Ration. Mech. Anal., 68 (1978), pp. 179–190.

Chaplygin, S. A., *A New Method of Approximate Integration of Differential Equations*, Moscow–Leningrad, 1950 [Russian].

Cholewa, J., *On certain non-typical properties of solutions of nonlinear elliptic equations*, Wiad. Mat., 38 (2002), pp. 53–60 [Polish].

Chow, S.-N. and Shen, W.-X., *Dynamics in a discrete Nagumo equation: Spatial topological chaos*, SIAM J. Appl. Math., 55 (1995), pp. 1764–1781.

čojczyk-Krolikiewicz, I., *Differential-functional inequalities of parabolic and elliptic type in bounded domain*, Zesz. Nauk. Pol. Ňlźskiej, Ser. matem.-fiz., 68 (1993), pp. 121–133.

Conway, E. and Smoller, J., *A comparison theorem for system of reaction-diffusion equations*, Commun. Part. Diff. Eq., 2 (1977), pp. 679–697.

Conway, E., Hoff, D. and Smoller, J., *Large time behavior of solutions of systems of nonlinear reaction-diffusion equations*, SIAM J. Appl. Math., 35 (1978), pp. 1–16.

Cronin, J., *Mathematics of Cell Electrophysiology*, Lecture Notes in Pure and Applied Mathematics, Vol. 63, Marcel Dekker, New York, 1981.

Czapiski, T., *Hyperbolic Functional Differential Equations*, Univ. of Gdańsk Publ., Gdańsk, 1999.

Deimling, K., *Ordinary Differential Equations in Banach Spaces*, Lecture Notes in Mathematics, Vol. 596, Springer-Verlag, Berlin, 1977.

DiBenedetto, E., *Partial Differential Equations*, Birkhäuser, Boston, 1995.

Diekmann, O. and Temme, N. M. (Eds.), *Nonlinear Diffusion Problems*, 2nd ed., MC Syllabus 28, Mathematisch Centrum, Amsterdam, 1982.

Dolgosheina, E. B., Karulin, A. Yu. and Bobylev, A. V., *A kinetic model of the agglutination process*, Math. Biosci., 109 (1992), pp. 1–10.

Drakhlin, M. and Litsyn, E., *Volterra operator: back to the future*, J. Nonlinear Convex Anal., 6 (2005), pp. 370–391.

Du, S. W. and Lakshmikantham, V., *Monotone iterative technique for differential equations in a Banach space*, J. Math. Anal. Appl., 87 (1982), pp. 454–459.

Eidelman, S. D., *Parabolic Systems*, North-Holland, 1969.

Ermentrout, G. B. and Terman, D. H., *Mathematical Foundations of Neuroscience*, Springer, New York, 2010.

Erneux, T. and Nicolis, G., *Propagating waves in discrete bistable reaction-diffusion systems*, Physica D, 67 (1993) 237–244.

Fath, G., *Propagation failure of travelling waves in a discrete bistable medium*, Physica D, 116 (1998), pp. 176–190.

Fife, P. C., *Mathematical Aspects of Reacting and Diffusing Systems*, Lecture Notes in Biomathematics, Vol. 28, Springer-Verlag, New York, 1979.

FitzHugh, R., *Mathematical models of excitation and propagation in nerves*. In: H. P. Schwann (Ed.), Biological Engineering, McGraw-Hill, New York, 1969.

Friedman, A., *Partial Differential Equations of Parabolic Type*, Prentice-Hall, Inc., Englewood Cliffs, New Jersey, 1964.

Friedman, A., *Partial Differential Equations of Parabolic Type*, Dover Publications. New York, 2008.

Gerstner, W. and Kistler, W. M., *Spiking Neuron Models*, Cambridge University Press, New York, 2002.

Gilbarg, D. and Trudinger, N. S., *Elliptic Partial Differential Equations of Second Order*, Springer-Verlag, Berlin, 2001.

Goltser, Y. and Litsyn, E., *Volterra integro-differential equations and infinite systems of ordinary differential equations*, Math. Comput. Modelling, 42 (2005), pp. 221–233.

Goltser, Y. and Litsyn, E., *Non-linear Volterra IDE, infinite systems and normal forms of ODE*, Nonlinear Anal., 68 (2008), pp. 1553–1569.

Granas, A. and Dugundji, J., *Fixed Point Theory*, Springer-Verlag, New York, 2003.

Grindrod, P. and Sleeman, B. D., *Homoclinic solutions for coupled systems of differential equations*, Proc. R. Soc. Edinb., 99 (1985), pp. 319–328.

Hastings, J. P., *The existence of homoclinic and periodic orbits for the Nagumo's equation*, Quart. J. Math., 27 (1976), pp. 123–134.

Hille, E. and Phillips, R. S., *Functional Analysis and Semi-groups*, American Mathematical Society, Providence, RI, 1957.

Hodgkin, A. L. and Huxley, A. F., *A quantitative description of membrane current and its applications to conduction and excitation in nerve*, J. Physiol., (Lond.), 117 (1952), pp. 500–544.

Jaruszewska-Walczak, D., *Generalized solutions of the Cauchy problem for infinite systems of functional differential equations*, Funct. Differ. Equ., 6 (1999), pp. 305–326.

Jaruszewska-Walczak, D., *Comparison theorem for infinite systems of parabolic functional-differential equations*, Ann. Polon. Math., 77 (2001), pp. 261–270.

Jaruszewska-Walczak, D., *Difference methods for infinite systems of hyperbolic functional differential equations on the Haar pyramid*, Opuscula Math., 24 (2004), pp. 85–96.

Jaruszewska-Walczak, D., *Infinite systems of hyperbolic differential-functional inequalities*, Univ. Iagel. Acta Math., 43 (2005), pp. 219–228.

Kamont, Z., *On the Chaplygin method for partial differential-functional equations of the first order*, Ann. Polon. Math., 38 (1980), pp. 27–46.

Kamont, Z., *Hyperbolic Functional Differential Inequalities and Applications*, Mathematics and its Applications, Vol. 486, Kluwer Academic Publishers, Dordrecht, 1999.

Kamont, Z., *Infinite systems of hyperbolic functional differential inequalities*, Nonlinear Anal., 51 (2002), pp. 1429–1445.

Kamont, Z., *Infinite systems of hyperbolic functional differential equations*, Ukrainian Math. J., 55 (2003), pp. 2006–2030.

Kamont, Z. and Kozieş, S., *Differential difference inequalities generated by infinite systems of quasilinear parabolic functional differential equations*, Funct. Differ. Equ., 10 (2003), pp. 215–238.

Kamont, Z. and Zacharek, S., *The line method for parabolic differential-functional equations with initial boundary conditions of the Dirichlet type*, Atti Sem. Mat. Fis. Univ. Modena, 35 (1987), pp. 249–262.

Kantorovič, L. and Akilov, G., *Functional Analysis*, Pergamon Press, Oxford, 1964.

Kastenberg, W. E., *Comparison theorems for nonlinear multicomponent diffusion systems*, J. Math. Anal. Appl., 29 (1970), pp. 299–304.

Keener, J. P., *Propagation and its failure in coupled systems of discrete-excitable cells*, SIAM J. Appl. Math., 47 (1987), pp. 556–572.

Keener, J. and Sneyd, J., *Mathematical Physiology*, Springer, New York, 1998.

Kellems, A. R., Chaturantabut, S., Sorensen, D. C. and Cox, S. J., *Morphologically accurate reduced order modeling of spiking neurons*, J. Comput. Neurosci., 28 (2010), pp. 477–494.

Keller, H. B., *Elliptic boundary value problems suggested by nonlinear diffusion processes*, Arch. Ration. Mech. Anal., 5 (1969), pp. 363–381.

Kozieş, S., *Differential difference inequalities generated by infinite systems of parabolic functional differential equations*, Commun. Math., 44 (2004), pp. 99–126.

Kraśnicka, B., *On some properties of solutions to a mixed problem for an infinite system of parabolic differential-functional equations in an unbounded domain*, Demonstratio Math., 15 (1982), pp. 229–240.

Kraśnicka, B., *On some properties of solutions to the first Fourier problem for infinite system of parabolic differential-functional equations in an arbitrary domain*, Univ. Iagel. Acta Math., 26 (1987), pp. 67–74.

Krasnosel'skiĭ, M. A., *Topological Methods in the Theory of Nonlinear Integral Equations*, Macmillan, New York, 1964.

Krasnosel'skiĭ, M. A., *Positive Solutions of Operator Equations*, Noordhoff, Groningen, 1964.

Kreĭn, M. G. and Rutman, M. A., *Linear operators for which a cone in a Banach space is invariant*, Uspekhi Mat. Nauk., 3 (1948), pp. 3–85 [Russian].

Kreyszig, E., *Introductory Functional Analysis with Applications*, Wiley & Sons, New York, 1978.

Krzyzański, M., *Évaluations des solutions de l'équation aux dérivées partielles du type parabolique, déterminées dans un domaine non borneé*, Ann. Polon. Math., 4 (1957), pp. 93–97.

Krzyzański, M., *Certain inéqualités relatives aux solutions de l'équation parabolique linéaire normale*, Bull. Acad. Polon. Sci. Math. Astr. Phys., 7 (1959), pp. 131–135.

Kusano, T., *On the Cauchy problem for a class of multicomponent diffusion systems*, Proc. Japan Acad., 39 (1963), pp. 634–638.

Kusano, T., *On the first boundary problem for quasilinear systems of parabolic differential equations in non-cylindrical domains*, Funkcial. Ekvac., 7 (1965), pp. 103–118.

Lachowicz, M. and Wrzosek, D., *A nonlocal coagulation-fragmentation model*, Appl. Math., 27 (2000), pp. 45–66.

Ladde, G. S., Lakshmikantham, V. and Vatsala, A. S., *Monotone Iterative Techniques for Nonlinear Differential Equations*, Pitman, Boston, 1985.

Ladyženskaja, O. A. and Ural'tseva, N. N., *Linear and Quasilinear Elliptic Equations*, Academic Press, New York, 1968.

Ladyženskaja, O. A., Solonnikov, V. A. and Ural'ceva, N. N., *Linear and Quasilinear Equations of Parabolic Type*, Mathematical Monographs, Vol. 23, Amer. Math. Soc., Providence, RI, 1968.

Lakshmikantham, V., *An extension of the method of quasilinearization*, J. Optim. Theory Appl., 82 (1994), pp. 315–321.

Lakshmikantham, V. and Drici, Z., *Positivity and boundedness of solutions of impulsive reaction-diffusion equations*, J. Comput. Appl. Math., 88 (1998), pp. 175–184.

Lakshmikantham, V. and Leela, S., *Differential and Integral Inequalities*, Vols. 1 and 2, Academic Press, New York, 1969.

Lakshmikantham, V. and Vatsala, A. S., *Generalized quasilinearizations versus Newton's method*, Appl. Math. Comput., 164 (2005), pp. 523–530.

Lamb, W., *Existence and uniqueness results for the continuous coagulation and fragmentation equation*, Math. Methods Appl. Sci., 27 (2004), pp. 703–721.

Laurençot, Ph., *On a class of conditions coagulation-fragmentation equation*, J. Differential Equations, 167 (2000), pp. 245–174.

Laurençot, Ph. and Wrzosek, D., *Fragmentation-diffusion model. Existence of solutions and their asymptotic behaviour*, Proc. Roy. Soc. Edinburgh Sect. A, 128 (1998), pp. 759–774.

Lemmert, R., *Existenzsätze für gewöhnliche Differentialgleichungen in geordneten Banachräumen*, Funkcial. Ekvac., 32 (1989), pp. 243–249.

Leszczyński, H., *On the method of lines for a heat nonlinear equation with functional dependence*, Ann. Polon. Math., 69 (1998a), pp. 61–74.

Leszczyński, H., *Parabolic Equations with Functional Dependence at Derivatives*, Univ. of Gdansk Publ., Gdańsk, 1998b.

Leszczyński, H., *Maximum principle and Chaplygin method for parabolic differential-functional systems*, Univ. Iagel. Acta Math., 37 (1999), pp. 283–300.

Leszczyński, H., *Quasilinearization methods for a nonlinear heat equation with functional dependence*, Georgian Math. J., 7 (2000), pp. 97–116.

Leszczyński, H., *Comparison ODE theorems related to the method of lines*, J. Appl. Anal., 17 (2011), pp. 137–154.

Leung, A., *Systems of Nonlinear Partial Differential Equations, Applications to Biology and Engineering*, Kluwer Academic Publishers, Dordrecht, 1989.

Leung, K. V., Mangeron, D., Oguztöreli, M. N. and Stein, R. B., *On the stability and numerical solutions of two neural models*, Util. Math., 5 (1974), pp. 167–212.

Lindsay, A. E., Lindsay, K. A. and Rosenberg, J. R., *Increased computational accuracy in multi-compartmental cable models by a novel approach for precise point process localization*. J. Comput. Neurosci., 19 (2005), pp. 21–38.

Lindsay, A. E., Lindsay, K. A. and Rosenberg, J. R., *New concepts in compartmental modeling*, Comput. Visual Sci., 10 (2007), pp. 79–98.

Liz, E., *Monotone iterative techniques in ordered Banach spaces*, Nonlinear Anal., 30 (1977), pp. 5179–5190.

Liz, E. and Nieto, J. J., *An abstract monotone iterative method and applications*, Dynam. Systems Appl., 7 (1998), pp. 365–376.

Logan, J. D., *An Introduction to Nonlinear Partial Differential Equations*, Wiley-Interscience, New York, 1994.

Lusin, N. N., *On the Chaplygin method of integration*, Collected Papers, Vol. 3, pp. 146–167, Moscow, 1953 [Russian].

Malec, M., *Sur une méthode des differences finies pour une équation non linéaire differentielle fonctionnelle aux dérivées mixtes*, Ann. Polon. Math., 36 (1979), pp. 1–10.

Maurin, K., *Methods of Hilbert spaces*, Polish Sci. Publ., Warszawa, 1959 [Polish].

McKean, H. P., *Nagumo's equation*, Adv. Math., 4 (1970), pp. 209–223.

McLaughlin, D. J., Lamb, W. and Bride, A. C., *Existence results for non-autonomous multiple-fragmentation models*, Math. Methods Appl. Sci., 20 (1997a), pp. 1313–1323.

McLaughlin, D. J., Lamb, W. and Bride, A. C., *A semigroup approach to fragmentation models*, SIAM J. Math. Anal., 28 (1997b), pp. 1158–1172.

McLaughlin, D. J., Lamb, W. and Bride, A. C., *An existence and uniqueness theorem for a coagulation and multiple-fragmentation equation*, SIAM J. Math. Anal., 28 (1997c), pp. 1173–1190.

McLaughlin, D. J., Lamb, W. and McBride, A. C., *Existence and uniqueness results for the non-autonomous coagulation and multiple-fragmentation equation*, Math. Methods Appl. Sci., 21 (1998), pp. 1067–1084.

McNabb, A., *Comparison and existence theorems for multicomponent diffusion systems*, J. Math. Anal. Appl., 3 (1961), pp. 133–144.

Mikhlin, S. G. and Smolickiĭ, H. L., *Approximate Methods of Differential and Integral Equations*, Nauka Press, Moscow, 1965 [Russian].

Mlak, W., *Differential inequalities of parabolic type*, Ann. Polon. Math., 3 (1957), pp. 349–354.

Mlak, W., *Parabolic differential inequalities and the Chaplygin's method*, Ann. Polon. Math., 8 (1960), pp. 139–153.

Mlak, W. and Olech, *Integration of infinite systems of differential inequalities*, Ann. Polon. Math., 13 (1963), pp. 105–112.

Moszyski, K. and Pokrzywa, A., *Sur les systémes infinis d'équations différentielles ordinaires dans certains espaces de Fréchet*, Polish Sci. Publ., Warszawa, 1974.

Mysovskikh, I. P., *Application of Chaplygin's method to the Dirichlet problem for elliptic equations of a special type*, Dokl. Akad. Nauk., 99 (1954), pp. 13–15 [Russian].

Nagumo, M., *Note in "Kansū-Hōteisiki"*, No. 15 (1939) [Japanese].

Nagumo, M. and Simoda, S., *Note sur l'inéqualité différentielle concernant les équations du type parabolique*, Proc. Japan Acad. Ser. A Math. Sci., 27 (1951), pp. 536–539.

Nagumo, J., Arimoto, S. and Yoshizawa, S., *An active pulse transmission line simulating nerve axon*, Proc. IRE, 50 (1962), pp. 2061–2070.

Nelson, M. I., *Comparison theorems for multicomponent diffusion systems: developments since 1961*, J. Appl. Math. Decis. Sci., 4 (2000), pp. 151–163.

Netka, M., *Differential difference inequalities related to parabolic functional differential equations*, Opuscula Math., 30 (2010), pp. 95–115.

Nickel, K., *Fehlerschranken und Eindeutigkeitsaussagen für die Lösungen nichtlinearer, stark gekoppelter parabolischer Differentialgleichungen*, Math. Z., 152 (1976), pp. 33–45.

Nickel, K., *Das Lemma von Max Müller-Nagumo-Westphal für stark gekoppelte Systeme parabolischer Functional-Differentialgleichungen*, Math. Z., 161 (1978), pp. 221–234.

Nickel, K., *Bounds for the set of solutions of functional-differential equations*, Ann. Polon. Math., 42 (1983), pp. 241–257.

Nowotarska, M., *Remark on the Chaplygin method for parabolic equations in unbounded domains*, Zesz. Nauk. UJ, Prace Matem., 17 (1975), pp. 115–117.

Oguztöreli, M. N., *On the neural equations of Cowan and Stein*, Util. Math., 2 (1972), pp. 305–315.

Omurtag, A. and Lytton, W., *Spectral method and high-order finite differences for the nonlinear cable equation*, Neural Comput., 22 (2010), pp. 2113–2136.

Pachpatte, B. G., *Inequalities for Differential and Integral Equations*, Academic Press, San Diego, 1998.

Pao, C. V., *Successive approximations of some nonlinear initial-boundary value problems*, SIAM J. Math. Anal., 5 (1974), pp. 91–102.

Pao, C. V., *Positive solutions of a nonlinear boundary-value problem of parabolic type*, J. Differential Equations, 22 (1976), pp. 145–163.

Pao, C. V., *Nonlinear Parabolic and Elliptic Equations*, Plenum Press, New York, 1992.

Pao, C. V., *Reaction diffusion equations with nonlocal boundary and nonlocal initial conditions*, J. Math. Anal. Appl., 195 (1995), pp. 702–718.

Pao, C. V., *Asymptotic behavior of solutions of reaction-diffusion equations with nonlocal boundary conditions*, J. Comput. Appl. Math., 88 (1998), pp. 225–238.

Pao, C. V., *Numerical analysis of coupled systems of nonlinear parabolic equations*, SIAM J. Numer. Anal., 36 (1999), pp. 393–416.

Pao, C. V., *Strongly coupled elliptic systems and applications to Lotka-Voltera models with cross-diffusion*, Nonlinear Anal., 60 (2005), pp. 1197–1217.

Pao, C. V., *Quasilinear parabolic and elliptic equations with nonlinear boundary conditions*, Nonlinear Anal., 66 (2007a), pp. 639–662.

Pao, C. V., *Numerical methods for quasi-linear elliptic equations with nonlinear boundary conditions*, SIAM J. Numer. Anal., 45 (2007b), pp. 1081–1106.

Pao, C. V. and Ruan, W. H., *Positive solutions of quasilinear parabolic systems with nonlinear boundary conditions*, J. Math. Anal. Appl., 333 (2007), pp. 472–499.

Pazy, A., *Semigroups of Linear Operators and Applications to Partial Differential Equations*, Springer, Berlin, 1983.

Pelczar, A., *On the method of successive approximations*, Polish Mathematical Society Symposium in Memory of T. Ważewski, Wiad. Mat., 20 (1976), pp. 80–84, [Polish].

Persidskiĭ, K. P., *Infinite countable systems of differential equations and stability of their solutions*, Part I, Izv. Akad. Nauk. Kaz. SSR, 7 (1958), pp. 52–71 [Russian].

Persidskiĭ, K. P., *Infinite countable systems of differential equations and stability of their solutions*, Part II, Izv. Akad. Nauk. Kaz. SSR, 8 (1959), pp. 45–64 [Russian].

Persidskiĭ, K. P., *Infinite countable systems of differential equations and stability of their solutions*, Part III, *Fundamental theorems on solvability of solutions of countable many differential equations*, Izv. Akad. Nauk. Kaz. SSR, 9 (1960), pp. 11–34 [Russian].

Persidskiĭ, K. P., *Selected Works*, Vol. 2, Nauka, 1976 [Russian].

Piotrowicz, W., *Investigation of infinite systems of differential equations*, Demonstratio Math., 21 (1988), pp. 1123–1137.

Plis, A., *The problem of uniqueness for the solution of a system of partial differential equations*, Bull. Acad. Polon. Sci. Cl. III, II (1954), pp. 55–57.

Pogorzelski, W., *Sur le systéme d'équations intégrales á une infinite de fonctions inconnues*, Ann. Polon. Math., 2 (1955), pp. 106–117.

Pogorzelski, W., *Propriétés des intégrals de l'équation parabolique normale*, Ann. Polon. Math., 4 (1957), pp. 61–92.

Pogorzelski, W., *Integral Equations and their Applications*, Vol. 1, Pergamon Press, Oxford, 1966.

Protter, M. H. and Weinberger, H. F., *Maximum Principles in Differential Equations*, Springer-Verlag, New York, 1984.

Prykarpatsky, A., Brzychczy, S., Samoylenko, V. K., *Finite-dimensional reductions of conservative dynamical systems and numerical analysis*, I, Ukrainian Math. J., 53 (2001), pp. 249–258.

Pudełko, A., *Existence and uniqueness of solutions of the Cauchy problem for nonlinear infinite systems of parabolic differential-functional equations*, Univ. Iagel. Acta Math., 40 (2002), pp. 49–56.

Pudełko, A., *Existence of solutions of the Cauchy problem for semilinear infinite systems of parabolic differential-functional equations*, Univ. Iagel. Acta Math., 42 (2004), pp. 149–169.

Pudełko, A., *Monotone iteration for infinite systems of parabolic equations*, Opuscula Math., 25 (2005a), pp. 307–318.

Pudełko, A., *Existence of solutions for infinite systems of parabolic equations with functional dependence*, Ann. Polon. Math., 86 (2005b), pp. 123–135.

Pudełko, A., *Monotone iteration for infinite systems of parabolic equations with functional dependence*, Ann. Polon. Math., 90 (2007), pp. 1–19.

Rabczuk, R., *Elements of Differential Inequalities*, Polish Sci. Publ., Warszawa, 1976 [Polish].

Rall, W., *Theoretical significance of dendritic trees for neuronal input-output relation*. In: R. F. Reiss (Ed.), Neural Theory and Modeling, Stanford University Press, Palo Alto, CA, 1964.

Redheffer, R. and Walter, W., *Existence theorems for strongly coupled systems of partial differential equations over Bernstein classes*, Bull. Amer. Math. Soc., 82 (1976), pp. 899–902.

Redheffer, R. and Walter, W., *Das Maximumprinzip in unbeschränkten Gebieten für parabolische Ungleichungen mit Funktionalen*, Math. Ann., 226 (1977), pp. 155–170.

Redheffer, R. and Walter, W., *Comparison theorems for parabolic functional inequalities*, Pacific J. Math., 82 (1979), pp. 447–470.

Redheffer, R. and Walter, W., *Stability of the null solution of parabolic functional inequalities*, Trans. Amer. Math. Soc., 262 (1980), pp. 285–302.

Redlinger, R., *Existence theorems for semilinear parabolic systems with functionals*, Nonlinear Anal., 8 (1984), pp. 667–682.

Redlinger, R., *On Volterra's population equation with diffusion*, SIAM J. Math. Anal., 16 (1985), pp. 135–142.

Redlinger, R., *Lower and upper solutions for strongly coupled systems of reaction-diffusion equations*. Proceedings of the International Conference on Theory and Applications of Differential Equations, Vol. II, Ohio University Press, Athens, 1988, pp. 327–332.

Rinzel, J., *Models in Neurobiology*. In: R. H. Enns, B. L. Jones, R. M. Miura and S. S. Rangnekar (Eds.), *Nonlinear Phenomena in Physics and Biology*, Plenum Press, New York and London, 1981, pp. 345–367.

Rothe, F., *Global Solutions of Reaction-Diffusion Systems*, Lecture Notes in Mathematics, Vol. 1072, Springer-Verlag, Berlin, 1984.

Rzepecki, B., *On infinite systems of differential equations with deviated argument*, Part I, Ann. Polon. Math., 31 (1975), pp. 159–169.

Rzepecki, B., *On infinite systems of differential equations with deviated argument*, Part II, Ann. Polon. Math., 34 (1977), pp. 251–264.

Sattinger, D. H., *Monotone methods in nonlinear elliptic and parabolic boundary value problems*, Indiana Univ. Math. J., 21 (1972), pp. 979–1000.

Sattinger, D. H., *Topics in Stability and Bifurcation Theory*, Lecture Notes in Mathematics, Vol. 309, Springer-Verlag, Berlin, 1973.

Schäfer, U., *An existence theorem for a parabolic differential equation in $\ell^\infty(A)$ based on the Tarski fixed point theorem*, Demonstratio Math., 30 (1997), pp. 461–464.

Scott, A. C., *Neuroscience: A Mathematical Primer*, Springer, Berlin and New York, 2002.

Scott, A. C., *Nonlinear Science Emergence and Dynamics of Coherent Structures*, 2nd ed., Oxford University Press, Oxford and New York, 2003.

Segev, I. and Burke, R. E., *Compartmental models of complex neurons*. In: C. Koch and I. Segev (Eds.), Methods in Neuronal Modeling, 2nd ed., MIT Press, 1998.

Smoller, J., *Shock Waves and Reaction-Diffusion Equations*, Springer-Verlag, New York, 1983.

Szarski, J., *Differential Inequalities*, Polish Sci. Publ., Warszawa, 1965.

Szarski, J., *Uniqueness of solutions of a mixed problem for parabolic differential-functional equations*, Ann. Polon. Math., 28 (1973), pp. 57–65.

Szarski, J., *Strong maximum principle for non-linear parabolic differential-functional inequalities*, Ann. Polon. Math., 49 (1974), pp. 207–214.

Szarski, J., *Strong maximum principle for nonlinear parabolic differential functional inequalities in arbitrary domains*, Ann. Polon. Math., 31 (1975), pp. 197–203.

Szarski, J., *Uniqueness of the solution to a mixed problem for parabolic functional-differential equations in arbitrary domains*, Bull. Acad. Polon. Sci. Math. Astr. Phys., 24 (1976), pp. 841–849.

Szarski, J., *Comparison theorem for infinite systems of parabolic differential-functional equations and strongly coupled infinite systems of parabolic equations*, Bull. Acad. Polon. Sci. Math. Astr. Phys., 27 (1979), pp. 739–846.

Szarski, J., *Infinite systems of parabolic differential-functional inequalities*, Bull. Acad. Polon. Sci. Math. Astr. Phys., 28 (1980a), pp. 477–481.

Szarski, J., *Comparison theorems for infinite systems of differential-functional equations and strongly coupled infinite systems of first order partial differential equations*, Rocky Mountain J. Math., 10 (1980b), pp. 239–246.

Tadeusiewicz, R. (Ed.), *Theoretical Neurocybernetics*, Warsaw University Press, Warszawa, 2009 [Polish].

Tam, K. K., *Construcion of upper and lower solutions for a problem in combustion theory*, J. Math. Anal. Appl., 69 (1979), pp. 131–145.

Tam, K. K. and Ng, K. Y. K., *Construction of upper and lower solutions for flow past a non-uniformly heated plate*, J. Math. Anal. Appl., 59 (1977), pp. 531–549.

Tarski, A., *A lattice-theoretical fixpoint theorem and its applications*, Pacific J. Math., 5 (1955), pp. 285–309.

Tuckwell, H. C., *Introduction to Theoretical Neurobiology*, Cambridge University Press, Cambridge, 1988.

Tychonov, A. N., *On an infinite system of differential equations*, Mat. Sb., 41 (1934), pp. 551–555 [German and Russian].

Tychonov, A. N., *Théorèmes d'unicité pour l'eáuation de la chaleur*, Mat. Sb., 42 (1935), pp. 199–216.

Ugowski, H., *On integro-differential equations of parabolic and elliptic type*, Ann. Polon. Math., 22 (1970), pp. 255–275.

Ugowski, H., *On integro-differential equations of parabolic type*, Ann. Polon. Math., 25 (1971), pp. 9–22.

Ugowski, H., *Some theorems on the estimate and existence of solutions of integro-differential equations of parabolic type*, Ann. Polon. Math., 25 (1972), pp. 311–323.

Ugowski, H., *On a certain non-linear initial-boundary value problem for integro-differential equations of parabolic type*, Ann. Polon. Math., 28 (1973), pp. 249–259.

Ursell, F., *Infinite systems of equations. The effect of truncation*, Quart. J. Mech. Appl. Math., 49 (1996), pp. 217–230.

Vaĭnberg, M. M., *Variational Methods for the Study of Nonlinear Operators*, Holden-Day Inc., San Francisco, 1964.

Valeev, K. G. and Zhautykov, O. A., *Infinite Systems of Differential Equations*, Nauka, 1974 [Russian].

Vatsala, A. S. and Yang, J., *Monotone iterative technique for semilinear elliptic systems*, Bound. Value Probl., 2 (2005), pp. 93–106.

Vatsala, A. S. and Yang, J., *Generalized quasilinearization method for reaction diffusion systems*, Nonlinear Stud., 13 (2006), pp. 53–72.

Voigt, A., *The method of lines for nonlinear parabolic equations with mixed derivatives*, Numer. Math., 32 (1979), pp. 197–207.

Wake, G. C., *On comparison theorems for multicomponent diffusion systems*, J. Math. Anal. Appl., 26 (1969), pp. 292–296.

Walter, W., *Differential and Integral Inequalities*, Springer, New York, 1970.

Walter, W., *Differential inequalities and maximum principles: theory, new methods and applications*, Nonlinear Anal., 30 (1997), pp. 4695–4711.

Walter, W., *The parabolic cauchy problem and quenching*, Dyn. Contin. Discrete Impuls. Syst. Ser. A Math. Anal., 8 (2001), pp. 99–119.

Walter, W., *Nonlinear parabolic differential equations and inequalities*, Discrete Contin. Dyn. Syst., 8 (2002), pp. 451–468.

Wang, J., *Monotone method for diffusion equations with nonlinear diffusion coefficients*, Nonlinear Anal., 34 (1998), pp. 113–142.

Wazewski, T., *Sur le probléme de Cauchy relatif á un systém d'équations aux deriveés partielles*, Ann. Soc. Polon. Math., 15 (1936), pp. 101–127.

Wazewski, T., *Sur une extension du procédé de I. Jungermann pour établir la convergences des approximations successives au cas des équations différentielles ordinaires*, Bull. Acad. Polon. Sci. Math. Astr. Phys., 8 (1960a), pp. 43–46.

Wazewski, T., *Sur une procédé de prouver la convergence des approximations successives sans utilisation des séries de comparaison*, Bull. Acad. Polon. Sci. Math. Astr. Phys., 8 (1960b), pp. 47–52.

Westphal, H., *Zur Abschätzung der Lösungen nichtlinearer parabolischer Differentialgleichungen*, Math. Z., 51 (1949), pp. 690–695.

Wloka, J., *Grundräume und verallgemeinerte Funktionen*, Lecture Notes in Mathematics, Vol. 82, Springer-Verlag, Berlin, 1969.

Wloka, J., *Funktionalanalysis und Anwendungen*, de Gruiter, Berlin, 1979, [German].

Wrzosek, D., *Existence of solutions for the discrete coagulation-fragmentation model with diffusion*, Topol. Methods Nonlinear Anal., 9 (1997), pp. 279–296.

Wrzosek, D., *On singular properties of solutions of Smoluchowski equation systems*, Wiad. Mat., 35 (1999), pp. 11–35 [Polish].

Wrzosek, D., *Mass-conserving solutions to the discrete coagulation-fragmentation model with diffusion*, Nonlinear Anal., 49 (2002), pp. 297–314.

Wrzosek, D., *Weak solutions to the Cauchy problem for the diffusive discrete coagulation-fragmentation system*, J. Math. Anal. Appl., 289 (2004), pp. 405–418.

Wu, J., *Theory and Applications of Partial Functional Differential Equations*, Springer-Verlag, New York, 1996.

Zabawa, T. S., *Existence of solutions of the Dirichlet problem for an infinite system of nonlinear differential-functional equations of elliptic type*, Opuscula Math., 25 (2005), pp. 333–343.

Zabawa, T. S., *Stability of infinite systems of nonlinear differential-functional equations of parabolic type*, Opuscula Math., 26 (2006), pp. 173–183.

Zeragia, P. K., *Using Chaplygin's method for solving fundamental boundary value problems for nonlinear partial differential equations of parabolic type*, Soobshch. Akad. Nauk GSSR, 17 (1956), pp. 103–109 [Russian].

Zeragia, P. K., *Boundary value problems for certain nonlinear equations of parabolic type*, Trudy Tbil. Mat. Inst., 24 (1957), pp. 195–221 [Russian].

Zeragia, P. K., *Chaplygin's method for nonlinear hyperbolic differential equations with boundary conditions*, Proc. Tbil. Gos. Univ., 100 (1965), pp. 145–154 [Georgian] (English summary).

Zeragia, P. K., *Chaplygin's method for some boundary value problem for a certain class of nonlinear equations of parabolic type*, Proc. Tbil. Univ., A 6–7 (1973), pp. 17–27 [Russian].

Zhautykov, O. A., *Infinite systems of differential equations and their applications*, Differ. Uravn., 1 (1965), pp. 162–170 [Russian].

Zinner, B., *Stability of traveling wavefronts for the discrete Nagumo equation*, SIAM J. Math. Anal., 22 (1991), pp. 1016–1020.

Zinner, B., *Existence of travelling wave front solutions for the discrete Nagumo equation*, J. Differential Equations, 96 (1992), pp. 1–27.

Index

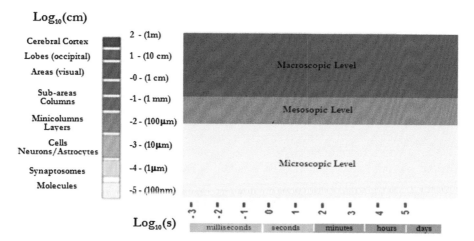

Figure 8.1 Illustrating the hierarchy of spatiotemporal evolution. Please see color plate at the back of the book.

Printed and bound by CPI Group (UK) Ltd, Croydon, CR0 4YY

08/05/2025

01865021-0001